Chapter 1: Introduction to Sports Analytics
- Overview of sports analytics
- The role of data in sports decision-making ... 5
- Historical perspective of sports analytics ... 7
- Importance of R in sports analytics ... 9

Chapter 2: Data Collection and Preparation ... 11
- Data sources in sports analytics ... 11
- Scraping data from websites and APIs ... 13
- Data cleaning and preprocessing ... 15
- Creating structured datasets for analysis ... 17

Chapter 3: Exploratory Data Analysis (EDA) ... 20
- Descriptive statistics in sports analytics ... 20
- Visualizing sports data with R ... 21
- Identifying trends and patterns in sports data ... 23
- EDA case studies from various sports ... 25

Chapter 4: Performance Metrics and Player Evaluation ... 28
- Common performance metrics in sports ... 28
- Creating custom metrics for player evaluation ... 29
- Comparative analysis of players and teams ... 31
- Player evaluation models and case studies ... 33

Chapter 5: Team Analysis and Strategy ... 38
- Team-level performance metrics ... 38
- Analyzing team dynamics and strategy ... 40
- Game simulation and strategy optimization ... 44
- Case studies of successful team strategies ... 48

Chapter 6: Injury Analysis and Prevention ... 51
- Injury data collection and analysis ... 51
- Identifying injury risk factors ... 55
- Injury prediction models ... 57
- Injury prevention strategies in sports ... 59

Chapter 7: Predictive Modeling in Sports ... 62
- Introduction to predictive modeling ... 62
- Building predictive models for sports outcomes ... 66
- Model evaluation and validation techniques ... 68
- Predicting game outcomes and player performance ... 73

Chapter 8: Player Draft and Recruitment ... 78
- Draft strategy in different sports ... 78
- Using data to evaluate draft prospects ... 80
- Case studies of successful draft selections ... 84
- Free agency and recruitment analytics ... 85

Chapter 9: Fan Engagement and Marketing — 90
- Analyzing fan behavior and preferences — 90
- Data-driven marketing strategies — 92
- Enhancing the fan experience through analytics — 96
- Case studies of successful fan engagement campaigns — 98

Chapter 10: Ethics and Fair Play in Sports Analytics — 101
- Ethical considerations in sports analytics — 101
- Fair play and data integrity — 103
- Privacy concerns and data security — 104
- Responsible use of analytics in sports — 106

Chapter 11: Case Studies in Sports Analytics — 109
- Real-world case studies from various sports — 109
- In-depth analysis of notable games or seasons — 110
- Lessons learned from successful analytics projects — 112
- Challenges and failures in sports analytics — 114

Chapter 12: Future Trends in Sports Analytics — 117
- Emerging technologies in sports analytics — 117
- The role of machine learning and AI — 119
- Predictions for the future of sports analytics — 120
- Opportunities and challenges in the field — 122

Chapter 13: Getting Started with R in Sports Analytics — 125
- Setting up an R environment for sports analytics — 125
- Basic R programming for data analysis — 127
- R packages and libraries for sports analytics — 129

Chapter 14: Advanced Topics in R for Sports Analytics — 132
- Advanced data manipulation in R for Sports Analytics — 132
- Building custom functions and packages for Sports Analytics in R — 134
- Integrating R with other tools and languages for Sports Analytics — 137
- Tips for efficient and scalable R coding for Sports Analytics — 139

Chapter 15: Conclusion and Practical Applications — 142
- Summarizing key takeaways from the book of Sports Analytics in R — 142
- Encouraging further exploration in sports analytics — 143
- Real-world applications of sports analytics using R — 145
- Final thoughts and future directions — 147

Example of Sports Analytics using R Code — 150
Advanced example of Soccer Analytics with R code — 152
Advanced example of NBA analytics with R code — 154
Advanced example of NFL analytics with R code — 156
Advanced example of Baseball analytics with R code — 158
Advanced example of Hockey analytics with R code — 160
Advanced example of Tennis analytics with R code — 162

Advanced example of Golf analytics with R code 164
Advanced example of Boxing analytics with R code 166

Chapter 1: Introduction to Sports Analytics

Overview of sports analytics

Sports analytics is a field of study that involves the collection, analysis, and interpretation of data related to sports. It has gained increasing prominence in recent years, transforming the way sports teams and organizations make decisions and improve their performance. Sports analytics leverages data to gain insights, make informed decisions, and gain a competitive edge in the world of sports. Here are some key aspects of sports analytics:

Data Collection: Sports analytics begins with the collection of vast amounts of data. This data can come from various sources, including player performance statistics, game footage, sensor technologies, and even social media. Advanced tools and technologies have made it easier to collect data in real-time, providing a wealth of information for analysis.

Performance Analysis: One of the primary applications of sports analytics is performance analysis. Coaches and analysts use data to evaluate player performance, team dynamics, and game strategies. They can identify strengths and weaknesses, helping teams make tactical decisions during games and practices.

Player Evaluation: Teams use analytics to evaluate and scout potential players. This involves assessing a player's skills, physical attributes, and potential contributions to the team. Data-driven player evaluation has become a crucial part of drafting and recruitment processes.

Injury Prevention: Analytics plays a role in injury prevention by analyzing injury data to identify risk factors and patterns. Teams can use this information to develop training programs and strategies to reduce the likelihood of injuries.

Fan Engagement: Sports organizations also use analytics to engage with fans. By understanding fan behavior and preferences, teams can tailor their marketing efforts, improve the in-stadium experience, and build a stronger fan base.

Game Strategy: Coaches use analytics to refine game strategies. This includes analyzing opponent data, simulating game scenarios, and optimizing plays. Analytics helps teams make data-driven decisions on when to use specific tactics during games.

Predictive Modeling: Predictive modeling is a key aspect of sports analytics. By building models using historical data, analysts can make predictions about future game outcomes, player performance, and even fan behavior. Predictive analytics assists teams in making strategic decisions.

Ethical Considerations: The use of data in sports analytics also raises ethical considerations, such as data privacy, fair play, and responsible use of analytics. Ethical guidelines are essential to ensure that data is used responsibly and transparently.

Technology and Tools: Advanced technologies, including wearable sensors, video analysis software, and data visualization tools, have revolutionized sports analytics. These tools enable real-time data collection and visualization, making it easier for teams to make decisions during games.

Continuous Evolution: Sports analytics is a dynamic field that continually evolves. New data sources, technologies, and analytical techniques emerge regularly, providing fresh opportunities for teams to gain a competitive advantage.

In summary, sports analytics has become an integral part of the sports industry, offering teams and organizations valuable insights to enhance performance, make informed decisions, and engage with fans. As the field continues to grow and evolve, it promises to shape the future of sports in exciting and innovative ways.

The role of data in sports decision-making

The role of data in sports decision-making is significant and has transformed the way teams, coaches, and organizations operate. Data-driven decision-making has become a fundamental aspect of modern sports, and here's an overview of its key roles:

Performance Analysis: Data is used to analyze player and team performance. Performance metrics, such as shooting accuracy, passing efficiency, and defensive stats, are collected and analyzed to identify strengths and weaknesses. Coaches use this information to make tactical decisions, devise game strategies, and determine lineups.

Injury Prevention: Injury data is crucial for understanding player health and reducing the risk of injuries. Sports organizations collect data on player injuries, including the type, frequency, and circumstances. This data helps in designing training programs that minimize injury risks and ensure player safety.

Player Evaluation and Recruitment: Teams use data to evaluate and recruit players. Scouts and analysts assess potential draft picks or free agents by examining their performance statistics, physical attributes, and other relevant data. Data-driven player evaluations play a pivotal role in roster decisions.

Game Strategy: Data-driven insights inform game strategy. Coaches analyze opponent data, historical game data, and situational statistics to make decisions during matches. They can adjust tactics based on real-time data and opponent behavior.

Fan Engagement: Data is used to engage with fans more effectively. Sports organizations collect data on fan behavior, preferences, and engagement across various channels. This information helps tailor marketing campaigns, promotions, and fan experiences to maximize fan engagement and loyalty.

Revenue Generation: Data can be a key driver of revenue generation. Teams and venues analyze ticket sales data, merchandise sales, and concessions data to optimize pricing strategies and maximize revenue streams.

Predictive Analytics: Predictive modeling uses historical data to make forecasts about future events. In sports, this can include predicting game outcomes, player performance, and even fan attendance. Predictive analytics guides strategic decisions, such as roster changes and marketing campaigns.

Training and Conditioning: Sports analytics provides insights into player conditioning and training. Wearable technology, such as fitness trackers and GPS devices, generates data on player movements, heart rate, and physical exertion. Coaches and sports scientists use this data to design personalized training programs that improve player fitness and performance.

Recruitment Scouting: Data analytics is used in recruitment scouting to identify talent at early stages. It helps in spotting promising young athletes who may have the potential to excel in a particular sport. Data-driven scouting allows teams to invest in the development of young talent.

Game Simulation: Teams use data-driven game simulations to strategize and make decisions. By simulating different game scenarios and strategies, teams can prepare for various situations and make informed choices during games.

Trend Analysis: Data helps sports organizations identify trends and patterns over time. This can include long-term player development trends, fan engagement trends, and the impact of rule changes on the game. Trend analysis assists in long-term planning and decision-making.

Competitive Advantage: Ultimately, data provides a competitive advantage. Teams that effectively collect, analyze, and leverage data are better positioned to make informed decisions, gain an edge over their competitors, and achieve success on the field or court.

In summary, data plays a multifaceted role in sports decision-making, impacting player performance, game strategies, fan engagement, revenue generation, and overall competitiveness. As technology continues to advance, the role of data in sports is likely to grow even more influential in the future.

Historical perspective of sports analytics

The historical perspective of sports analytics highlights the evolution of using data and statistical analysis in sports decision-making. While modern sports analytics has gained prominence in recent years, its roots can be traced back to several key developments and milestones:

Early Statistical Analysis (19th Century):
- The use of basic statistics, such as batting averages in baseball, dates back to the 19th century. These early statistics provided rudimentary insights into player performance.

Bill James and Sabermetrics (Late 20th Century):
- Bill James, a baseball enthusiast, is often credited with popularizing the term "sabermetrics." In the 1970s and 1980s, he began publishing books and articles that introduced more advanced statistical metrics to evaluate baseball players. Sabermetrics challenged conventional wisdom and led to the development of new statistics like OPS (On-base Plus Slugging) and WAR (Wins Above Replacement). It marked a significant shift towards a more data-driven approach in baseball.

Moneyball (Early 21st Century):
- The book "Moneyball" by Michael Lewis, published in 2003 and later adapted into a movie, showcased the transformative power of analytics in baseball. It highlighted how the Oakland Athletics, led by general

manager Billy Beane, used data and statistical analysis to assemble a competitive team despite a limited budget. Moneyball brought widespread attention to the impact of analytics in sports.

Technology Advancements (2000s - Present):
- The advent of advanced technology, including player tracking systems, wearable sensors, and high-speed cameras, revolutionized data collection in sports. These technologies provided rich datasets for performance analysis, enabling teams to track player movements, physiological data, and ball trajectories with high precision.

Sports Analytics Conferences and Research (2000s - Present):
- The growth of sports analytics conferences, such as the MIT Sloan Sports Analytics Conference, provided a platform for researchers, analysts, and sports professionals to share insights and best practices. Academic institutions and think tanks also contributed to the development of sports analytics as a field of study.

Expansion Beyond Baseball (2000s - Present):
- While baseball was an early adopter of analytics, other sports began to embrace data-driven decision-making. Basketball, football (soccer), and American football adopted analytics in areas like player tracking, shot analysis, and injury prevention. The NBA, for instance, introduced player tracking technology, which records player movement data at a high frequency.

Integration of Machine Learning and AI (2010s - Present):
- In recent years, machine learning and artificial intelligence have played a growing role in sports analytics. These technologies are used for predictive modeling, player performance analysis, and game strategy optimization. Machine learning algorithms can uncover complex patterns and insights from large datasets.

Global Reach and Influence (2010s - Present):
- The influence of sports analytics has expanded globally, with teams and organizations from various sports and countries adopting data-driven approaches. International competitions, such as the FIFA World Cup and the Olympic Games, have also integrated analytics into their preparations and strategies.

Ethical Considerations and Data Privacy (2010s - Present):
- As sports analytics has advanced, ethical considerations surrounding data privacy, player consent, and fair play have come to the forefront.

Organizations and governing bodies have had to establish guidelines to ensure responsible use of data in sports.

The historical perspective of sports analytics reflects a journey from basic statistics to advanced data-driven decision-making. It demonstrates how analytics has become an integral part of the sports industry, impacting player evaluation, game strategy, injury prevention, and fan engagement across various sports. The field continues to evolve, driven by technological advancements and a growing understanding of the potential of data in sports.

Importance of R in sports analytics

R is a powerful and versatile programming language and environment for statistical analysis and data visualization. Its importance in sports analytics is significant due to several key advantages it offers:

Statistical Analysis: R provides a wide range of statistical functions and libraries that are essential for analyzing sports data. Analysts can perform descriptive statistics, hypothesis testing, regression analysis, and more to gain insights into player performance, team dynamics, and game outcomes.

Data Visualization: R excels at data visualization. It offers libraries like ggplot2 for creating informative and visually appealing plots and charts. Visualizations are crucial for presenting findings, identifying trends, and communicating insights to coaches, players, and stakeholders.

Data Manipulation: R offers extensive data manipulation capabilities through libraries like dplyr and tidyr. Analysts can clean, transform, and reshape data to prepare it for analysis efficiently. This is particularly important when dealing with messy sports data from various sources.

Integration with Data Sources: R supports data integration from various sources, including databases, CSV files, APIs, and web scraping. This flexibility allows sports analysts to collect and aggregate data from different platforms and systems.

Machine Learning: R has a robust ecosystem for machine learning with packages like caret, randomForest, and xgboost. This is crucial for building predictive models, player evaluation, and game outcome prediction, enhancing decision-making in sports.

Customization: R allows for custom function and package development. Analysts can create specialized tools and functions tailored to the specific needs of sports analytics, facilitating more efficient workflows and analysis.

Reproducibility: In sports analytics, it's important to ensure that analyses are reproducible. R's script-based approach makes it easy to document and reproduce analyses, which is crucial for collaboration and transparency within sports organizations.

Community and Resources: R has a large and active community of users and developers. This means there is a wealth of resources, tutorials, and packages available for sports analysts to tap into. It's easier to find solutions to specific challenges within this community.

Cost-Effective: R is open-source and free to use, making it a cost-effective choice for sports organizations, particularly those with limited budgets. This accessibility allows smaller teams and organizations to leverage data analytics for a competitive advantage.

Cross-Industry Applicability: R is not limited to sports analytics; it is used in various industries, including finance, healthcare, and marketing. This cross-industry applicability means that analysts with R skills can apply their knowledge to other domains if needed.

Flexibility in Reporting: R can generate dynamic and customizable reports and dashboards using tools like R Markdown and Shiny. This feature is useful for creating interactive reports that coaches and stakeholders can use to make informed decisions.

R plays a crucial role in sports analytics due to its extensive statistical capabilities, data visualization tools, flexibility, and affordability. It empowers sports analysts to extract meaningful insights from complex data, develop predictive models, and enhance decision-making processes within sports organizations. Its versatility and strong community support make it a valuable asset in the field of sports analytics.

Chapter 2: Data Collection and Preparation

Data sources in sports analytics

Data sources in sports analytics encompass a wide range of information that analysts and teams use to gain insights, make informed decisions, and improve performance. These sources provide a wealth of data on players, teams, and games. Here are some of the primary data sources in sports analytics:

Game Statistics:
- Official game statistics are a fundamental source of data in sports analytics. These statistics include information on goals, assists, rebounds, points scored, fouls, turnovers, and many other in-game events. Leagues and governing bodies often maintain comprehensive databases of these statistics.

Player Performance Data:
- Player-specific data includes individual performance metrics such as shooting accuracy, passing efficiency, tackles, steals, and more. This data helps in assessing player skills and contributions to the team.

Player Tracking Data:
- Player tracking systems use sensors and cameras to collect real-time data on player movements. This includes data on player positioning, speed, acceleration, distance covered, and player-to-player interactions. It is invaluable for tactical analysis and assessing player fitness.

Video Footage:
- Video analysis is a critical data source in sports analytics. Analysts review game footage to break down plays, study opponent strategies, and evaluate player performance. This data source provides context and qualitative insights.

Injury Data:
- Injury data tracks when, where, and how injuries occur during games and practices. This information is crucial for injury prevention and player safety.

Biometric Data:
- Wearable sensors and devices provide biometric data, including heart rate, body temperature, and fatigue levels. This data helps monitor player health and performance, allowing teams to optimize training programs.

Scouting Reports:
- Scouting reports contain data on potential draft picks, free agents, and opponents. Analysts assess player strengths, weaknesses, and suitability for the team based on these reports.

Social Media Data:
- Social media platforms generate data on fan engagement, sentiment analysis, and player interactions. Teams use this data to gauge fan sentiment and tailor marketing efforts.

Ticketing and Attendance Data:
- Ticket sales and attendance data provide insights into fan behavior and preferences. Teams use this data to optimize pricing, promotions, and fan experiences.

Weather Data:
- Weather conditions can significantly impact outdoor sports. Teams and venues collect weather data to prepare for adverse conditions and make adjustments as needed.

Financial Data:
- Financial data includes revenue, salary cap information, and budgeting data. It is essential for managing team finances, making roster decisions, and complying with league regulations.

Web and Social Media Analytics:
- Teams monitor website and social media analytics to track online engagement, website traffic, and fan interactions. This data helps in digital marketing and content strategy.

Coach and Player Interviews:
- Interviews with coaches and players can provide qualitative insights into team dynamics, strategies, and performance. These insights complement quantitative data analysis.

Publicly Available Data:
- Publicly available datasets, such as historical game scores and statistics, are often used for research and analysis by sports enthusiasts, analysts, and researchers.

Custom Data Sources:
- Sports organizations may develop custom data sources and applications to collect specific information relevant to their needs. This could include proprietary data on training regimens, nutrition, and more.

The availability of these data sources varies by sport and organization. In recent years, advancements in technology have led to an increase in the variety and depth of data available for sports analytics, contributing to a more data-driven approach in the world of sports.

Scraping data from websites and APIs

Scraping data from websites and APIs is a common practice in sports analytics to gather valuable information for analysis and decision-making. Here's an overview of how data scraping can be done in sports analytics:

1. Web Scraping:
Web scraping involves extracting data from websites. In sports analytics, web scraping can be used to collect data on player statistics, game scores, schedules, and more. Here's how it can be done:

a. Select a Programming Language:
- Choose a programming language suitable for web scraping. Popular choices include Python (with libraries like BeautifulSoup and Scrapy), R (with libraries like rvest), and JavaScript (with tools like Puppeteer).

b. Identify the Data Source:
- Determine the websites from which you want to scrape data. Ensure that the website's terms of service allow for web scraping, and be respectful of their policies.

c. Inspect Website Structure:
- Use browser developer tools to inspect the website's HTML structure. This will help you identify the elements that contain the data you need.

d. Write Web Scraping Code:
- Write code to access the website, retrieve HTML content, and parse it to extract the desired data. This may involve using CSS selectors or XPath expressions to locate specific elements.

e. Handle Pagination and Dynamic Content:
- Some websites paginate data or load it dynamically through JavaScript. Ensure your web scraping code can handle these scenarios.

f. Implement Error Handling:
- Include error-handling mechanisms in your code to handle situations where the website structure changes or the server rate-limits requests.

g. Store Data:
- After scraping data, store it in a structured format such as CSV, JSON, or a database for further analysis.

2. API Data Retrieval:
Many sports organizations and data providers offer APIs (Application Programming Interfaces) that allow developers to access sports-related data programmatically. APIs are typically more reliable and efficient than web scraping. Here's how you can retrieve data from APIs:

a. Register for API Access:
- If the sports organization or data provider offers an API, you may need to sign up and obtain an API key or access credentials.

b. Choose a Programming Language:
- Select a programming language that supports making HTTP requests. Python, R, and JavaScript are commonly used for this purpose.

c. Review API Documentation:
- Study the API documentation provided by the sports organization or data provider. It will detail the available endpoints, request parameters, and response formats.

d. Make API Requests:
- Write code to make HTTP requests to the API endpoints using the appropriate authentication and parameters. This could include retrieving player statistics, game data, or other relevant information.

e. Parse API Responses:
- Parse the JSON or XML responses returned by the API to extract the required data.

f. Handle Rate Limiting:
- Be aware of any rate-limiting policies imposed by the API provider. Implement rate-limiting and retry logic in your code if necessary.

g. Store Data:
- Store the API data in a structured format for analysis, such as a database or data file.

When scraping data from websites or using APIs in sports analytics, it's essential to stay updated with any changes in the data sources or APIs to ensure the reliability of your data collection processes. Additionally, always respect the terms of service and policies of the websites and API providers you interact with.

Data cleaning and preprocessing

Data cleaning and preprocessing are crucial steps in sports analytics (and data analysis in general) to ensure that the data you work with is accurate, reliable, and ready for analysis. Here are the key steps involved in data cleaning and preprocessing:

1. Data Collection:
   - Gather data from various sources, including websites, APIs, databases, or internal records.

2. Data Inspection:
   - Examine the raw data to understand its structure, format, and potential issues. Pay attention to missing values, outliers, and inconsistencies.

3. Handling Missing Data:
   - Identify missing data points and decide on an appropriate strategy:
     - Imputation: Replace missing values with estimated or calculated values (e.g., mean, median, or mode).
     - Removal: Remove rows or columns with a high percentage of missing data if they don't contain critical information.
     - Interpolation: Use techniques like linear interpolation for time-series data.

4. Dealing with Duplicates:
   - Detect and remove duplicate rows from the dataset, if applicable.

5. Outlier Detection and Handling:
   - Identify outliers using statistical methods or visualization techniques.
   - Decide whether to remove outliers or transform them, depending on the context of your analysis.

6. Data Conversion:
   - Ensure that data types are appropriate for analysis. Convert data types (e.g., converting strings to numerical values) as needed.

7. Standardization and Scaling:
   - Normalize or scale numerical features to a consistent range. Common techniques include Min-Max scaling and Z-score normalization.

8. Encoding Categorical Data:
   - Convert categorical variables (e.g., team names, player positions) into numerical format using techniques like one-hot encoding or label encoding.

9. Feature Selection:
   - Identify and select relevant features (columns) for analysis while excluding irrelevant or redundant ones. Feature selection techniques include correlation analysis and feature importance from machine learning models.

10. Handling Time-Series Data: - If working with time-series data (e.g., game scores over time), ensure it is properly sorted and handle any irregularities in time intervals.

11. Data Transformation: - Apply transformations to the data as needed, such as logarithmic transformations for skewed data or differencing for time-series data to stabilize variance.

12. Data Integration: - If you have multiple datasets, integrate them by merging or joining based on common identifiers (e.g., player IDs or game IDs).

13. Data Validation: - Validate the cleaned and preprocessed data to ensure it meets quality standards and does not contain errors or inconsistencies.

14. Exploratory Data Analysis (EDA): - Conduct exploratory data analysis to gain insights into the data's distribution, relationships, and patterns. Visualization tools and techniques are valuable for this step.

15. Documentation: - Keep records of the data cleaning and preprocessing steps performed, as well as any decisions made during the process. This documentation is crucial for transparency and reproducibility.

16. Data Splitting: - If you are building predictive models, split the dataset into training and testing sets to assess model performance accurately.

17. Handling Imbalanced Data (if applicable): - If your dataset is imbalanced (e.g., more wins than losses in a sports dataset), consider techniques like oversampling or undersampling to balance the classes.

18. Data Storage: - Store the cleaned and preprocessed data in a structured format, such as CSV, Excel, or a database, for easy access and analysis.

Data cleaning and preprocessing are iterative processes, and it's essential to revisit these steps as needed throughout the analysis project. Ensuring data quality and consistency is vital for obtaining meaningful insights and making informed decisions in sports analytics.

Creating structured datasets for analysis

Creating structured datasets for analysis is a crucial step in sports analytics, as it involves organizing and preparing data in a way that facilitates meaningful analysis and modeling. Here are the key steps to create structured datasets:

1. Data Collection:
   - Gather data from various sources, including websites, APIs, databases, or internal records, and store it in a raw, unprocessed format.

2. Data Cleaning and Preprocessing:
   - Perform data cleaning and preprocessing as described in the previous response. This step ensures that the data is free from errors, inconsistencies, and missing values.

3. Define the Analysis Goals:
   - Clearly define the objectives of your analysis. Determine what specific questions or problems you aim to address using the dataset.

4. Select Relevant Features:
   - Identify and select the features (columns) that are relevant to your analysis goals. Exclude irrelevant or redundant features to reduce noise and complexity.

5. Create a Data Dictionary:
    - Document the meaning and data type of each feature in a data dictionary. This documentation is essential for understanding the dataset and communicating it to others.

6. Establish Data Structures:
    - Decide on the appropriate data structures for your dataset. Common data structures include dataframes (e.g., in R or Python), tables (e.g., in SQL databases), or spreadsheets.

7. Data Integration (if applicable):
    - If you have multiple datasets from different sources, integrate them by merging or joining based on common identifiers (e.g., player IDs, game IDs).

8. Handle Time-Series Data (if applicable):
    - For time-series data, ensure that the timestamps are properly formatted and sorted. You may need to aggregate or resample data at specific time intervals.

9. Encode Categorical Data:
    - Convert categorical variables (e.g., team names, player positions) into numerical format using techniques like one-hot encoding or label encoding.

10. Data Transformation (if applicable): - Apply transformations to the data as needed, such as logarithmic transformations for skewed data or differencing for time-series data to stabilize variance.

11. Standardization and Scaling (if applicable): - Normalize or scale numerical features to a consistent range if your analysis or modeling techniques require it.

12. Handle Missing Data: - Ensure that any missing data has been appropriately handled through imputation or removal, as discussed in the previous response.

13. Create Target Variables (if applicable): - Define the target variable(s) you want to predict or analyze. For example, if you're predicting game outcomes, create a target variable indicating wins or losses.

14. Data Splitting (if applicable): - If you are building predictive models, split the dataset into training, validation, and testing sets. The proportions may vary depending on your analysis goals.

15. Data Validation: - Validate the structured dataset to ensure that it meets the requirements for your analysis goals and that there are no errors or inconsistencies.

16. Document the Data Processing Steps: - Keep detailed records of the data processing steps performed, including the rationale for each step. This documentation is essential for transparency and reproducibility.

17. Store the Structured Dataset: - Store the structured dataset in a format suitable for analysis, such as CSV, Excel, or a database, ensuring that it is well-organized and labeled.

Creating structured datasets is a critical foundation for any sports analytics project. Well-structured data simplifies the analysis process, enhances data integrity, and allows for more accurate modeling and decision-making in the field of sports analytics.

Chapter 3: Exploratory Data Analysis (EDA)

Descriptive statistics in sports analytics

Descriptive statistics play a vital role in sports analytics by providing a concise summary of key characteristics and patterns within the data. These statistics help analysts and sports professionals better understand and interpret various aspects of player performance, team dynamics, and game outcomes. Here are some common descriptive statistics used in sports analytics:

1. Measures of Central Tendency:
   - Mean (Average): The sum of all values divided by the number of data points. In sports analytics, the mean can be used to describe average player performance, team scoring, or other continuous variables.
   - Median: The middle value in a dataset when it is sorted in ascending or descending order. The median is less sensitive to extreme outliers than the mean and can provide a more robust measure of central tendency.
   - Mode: The most frequently occurring value in a dataset. In sports, the mode might be used to identify the most common player position or game outcome.

2. Measures of Dispersion:
   - Range: The difference between the maximum and minimum values in a dataset. It provides a sense of the spread of data and can be used to assess variability in player performance metrics.
   - Variance: A measure of how much data points deviate from the mean. It quantifies the spread or dispersion of data and is useful for understanding variability in statistics like player scoring.
   - Standard Deviation: The square root of the variance. It provides a measure of the average distance between data points and the mean. A higher standard deviation indicates greater variability.

3. Measures of Distribution:
   - Percentiles: Percentiles divide a dataset into 100 equal parts. Common percentiles include the 25th (first quartile), 50th (median), and 75th (third quartile) percentiles. They help understand the distribution of data and identify outliers.
   - Histograms: Histograms are graphical representations of the distribution of data. They provide a visual depiction of how data is distributed across different values or intervals.

- Box Plots: Box plots display the median, quartiles, and potential outliers in a dataset. They are helpful for comparing distributions, such as comparing the scoring distributions of two teams.

4. Frequency and Count Statistics:
   - Counts: Count statistics show the number of occurrences of specific events, such as goals scored by a player or wins by a team. They provide valuable insights into player and team performance.
   - Frequency Tables: These tables show the frequency of different categories or values within a dataset. In sports analytics, they can be used to display the frequency of player positions, game outcomes, or injury types.

5. Correlation Analysis:
   - Correlation Coefficient: Measures the strength and direction of the linear relationship between two continuous variables. In sports analytics, it can be used to assess the relationship between player statistics, such as the correlation between assists and goals.

6. Summary Statistics:
   - Summary Tables: Create summary tables that include key descriptive statistics for different aspects of sports data, such as player performance metrics, team statistics, and game outcomes.

Descriptive statistics help sports analysts and professionals summarize and interpret large datasets, making it easier to identify trends, outliers, and areas of interest. These statistics are the foundation for more advanced analyses, such as predictive modeling and hypothesis testing, in the field of sports analytics.

Visualizing sports data with R

Visualizing sports data with R can be a powerful way to gain insights, communicate findings, and engage stakeholders in the world of sports analytics. R offers various libraries and tools for creating a wide range of visualizations. Here are some common types of sports data visualizations you can create using R:

1. Scatter Plots:
   - Scatter plots are useful for displaying relationships between two numerical variables. In sports analytics, you can use them to visualize the relationship between player statistics, such as goals scored and assists.

 # Example scatter plot in R
plot(data$goals, data$assists, xlab="Goals", ylab="Assists", main="Goals vs. Assists")

2. Bar Charts:
   - Bar charts are effective for comparing categorical data or showing the distribution of data. They can be used to visualize team rankings, player positions, or game outcomes.

 # Example bar chart in R
barplot(table(data$team), main="Team Distribution", xlab="Team", ylab="Frequency")

3. Line Charts:
   - Line charts are suitable for displaying trends and changes over time. You can use them to visualize time-series data, such as game scores or player performance over multiple seasons.

 # Example line chart in R
plot(data$year, data$score, type="l", xlab="Year", ylab="Score", main="Game Score Over Time")

4. Box Plots:
   - Box plots are helpful for showing the distribution of data, including measures of central tendency and spread. They are useful for comparing player statistics or team performance.

 # Example box plot in R
boxplot(data$points, data$team, xlab="Team", ylab="Points", main="Points Distribution by Team")

5. Heatmaps:
   - Heatmaps can visualize patterns and relationships in data, such as game schedules or player performance matrices. They are useful for identifying clusters and trends.

 # Example heatmap in R
heatmap(matrix(data$performance, nrow=n, ncol=m), xlab="Players", ylab="Games", main="Player Performance Heatmap")

6. Pie Charts:
   - Pie charts are suitable for displaying the distribution of categorical data. You can use them to show the distribution of player positions or game outcomes.

 # Example pie chart in R
pie(table(data$position), main="Player Position Distribution")

7. Geographic Maps:
   - If your sports data includes geographical information, you can create maps to visualize player or team locations, venue distribution, or regional performance differences using packages like 'ggmap' or 'leaflet'.

 # Example geographic map in R using ggmap
library(ggmap)
map <- get_map(location="Your_Location", zoom=10)
ggmap(map)

8. Interactive Dashboards:
   - You can create interactive dashboards with R using packages like 'Shiny' to allow users to explore sports data dynamically, filter information, and view visualizations in real-time.

These are just a few examples of the types of sports data visualizations you can create with R. The choice of visualization depends on your specific analysis goals and the nature of the data you are working with. R's flexibility and a rich ecosystem of visualization libraries make it a versatile tool for sports analytics visualization.

Identifying trends and patterns in sports data

Identifying trends and patterns in sports data is a critical aspect of sports analytics. By uncovering these trends and patterns, sports analysts and professionals can make informed decisions, develop strategies, and gain insights into player performance, team dynamics, and game outcomes. Here's how you can go about identifying trends and patterns in sports data:

1. Data Exploration:
   - Begin by exploring the dataset to get a sense of its contents. Examine variables, data types, and summary statistics. Use tools like histograms, box plots, and summary tables to understand the distribution and central tendency of the data.

2. Time-Series Analysis:
   - If your data includes time-related information (e.g., game scores over seasons), perform time-series analysis to identify trends and seasonality. Use line charts or decompose time series to visualize patterns over time.

3. Correlation Analysis:
   - Conduct correlation analysis to determine relationships between variables. Calculate correlation coefficients and use scatter plots to visualize correlations. Identify strong positive or negative correlations that may reveal important connections.

4. Data Visualization:
   - Create various data visualizations, including line charts, bar charts, scatter plots, and heatmaps, to visually represent the data. Visualization can make trends and patterns more apparent.

5. Moving Averages:
   - Calculate moving averages to smooth out noise in time-series data. Moving averages can help reveal underlying trends by showing average values over a specific time window.

6. Clustering and Segmentation:
   - Apply clustering algorithms (e.g., k-means) to segment players, teams, or games into groups based on similarities in performance. This can uncover patterns within different clusters.

7. Hypothesis Testing:
   - Formulate hypotheses about potential trends or patterns and conduct hypothesis tests to confirm or reject them. Common statistical tests include t-tests, ANOVA, and chi-square tests.

8. Machine Learning:
   - Utilize machine learning algorithms, such as regression, decision trees, or neural networks, to predict outcomes and discover underlying patterns. Feature importance analysis can help identify the most influential variables.

9. Anomaly Detection:
   - Implement anomaly detection techniques to identify unusual or unexpected events, which can often indicate interesting patterns or outliers in the data.

10. Geographic Analysis:
- If your data includes geographical information (e.g., player locations or game venues), use geographic analysis to uncover spatial patterns or regional differences in performance.

11. Seasonal Decomposition:
- Decompose time-series data into its components (trend, seasonality, and residual) using methods like seasonal decomposition of time series (STL). This can reveal underlying patterns in data.

12. Natural Language Processing (NLP):
- If your data includes text data (e.g., player interviews or fan sentiment), use NLP techniques to extract insights, sentiments, or trends from textual information.

13. Dimensionality Reduction:
- Apply dimensionality reduction techniques like Principal Component Analysis (PCA) or t-SNE to reduce high-dimensional data to a lower dimension for pattern identification and visualization.

14. Expert Knowledge:
- Collaborate with domain experts, coaches, or sports analysts who have specialized knowledge in the sport. Their insights can guide the search for meaningful patterns and trends.

15. Time Series Forecasting:
- If your data involves time-series elements, use forecasting methods (e.g., ARIMA, exponential smoothing) to predict future trends based on historical patterns.

16. Social Network Analysis (SNA):
- Analyze social network data among players, teams, or fans to uncover network-based patterns and identify influential nodes or communities.

Identifying trends and patterns in sports data is an iterative process that involves data exploration, statistical analysis, visualization, and domain expertise. By combining these techniques, sports analysts can extract valuable insights that contribute to improved decision-making, strategy development, and overall performance in the sports industry.

EDA case studies from various sports

Exploratory Data Analysis (EDA) case studies from various sports can provide valuable insights into how data analysis techniques are applied to real-world sports datasets. Here are a few EDA case studies from different sports:

1. Baseball: Player Performance Analysis
   - Objective: Analyze the batting performance of baseball players in the Major League Baseball (MLB) using historical data.
   - Methods:
     - Data cleaning and preprocessing
     - Descriptive statistics and visualization of batting averages, home runs, and RBIs.
     - Comparison of performance between different teams and positions.

- Time-series analysis of player performance over seasons.
- Key Findings: Identify trends in player performance over time, differences between teams, and the impact of player positions on batting statistics.

2. Soccer: Team Performance Analysis
    - Objective: Evaluate the performance of soccer teams in a specific league over several seasons.
    - Methods:
        - Data cleaning and preprocessing
        - Calculation of various team performance metrics (e.g., points, goals scored, goals conceded).
        - Visualization of team rankings, goal differentials, and home vs. away performance.
        - Time-series analysis of team rankings and performance.
    - Key Findings: Identify top-performing teams, trends in team rankings, and home-field advantage effects on performance.

3. Basketball: Player Efficiency Analysis
    - Objective: Analyze the efficiency of NBA basketball players based on player statistics.
    - Methods:
        - Data cleaning and preprocessing
        - Calculation of player efficiency rating (PER) and other advanced metrics.
        - Visualization of player rankings, PER distribution, and correlations with team success.
        - Identification of outliers and top-performing players.
    - Key Findings: Discover which players have the highest PER, how PER relates to team success, and which statistics contribute most to player efficiency.

4. Tennis: Player Match Analysis
    - Objective: Analyze the performance of tennis players in Grand Slam tournaments.
    - Methods:
        - Data cleaning and preprocessing
        - Visualization of player rankings, match outcomes, and match durations.
        - Comparison of performance between male and female players.
        - Identification of player dominance periods.
    - Key Findings: Identify trends in Grand Slam winners, match durations, and gender differences in player performance.

5. American Football: Team Offensive Analysis
    - Objective: Analyze the offensive performance of NFL football teams in a particular season.

- Methods:
    - Data cleaning and preprocessing
    - Calculation of offensive statistics (e.g., yards gained, points scored).
    - Visualization of team rankings, offensive efficiency, and passing vs. rushing performance.
    - Analysis of red zone efficiency.
- Key Findings: Identify top offensive teams, trends in offensive performance, and factors contributing to red zone success.

These case studies showcase how EDA techniques can be applied to different sports datasets to uncover insights, trends, and patterns. Each sport presents unique challenges and opportunities for analysis, and EDA serves as the foundation for more advanced sports analytics tasks, such as predictive modeling and strategic decision-making.

Chapter 4: Performance Metrics and Player Evaluation

Common performance metrics in sports

Common performance metrics in sports vary depending on the specific sport and the position or role of the athlete. However, there are several general performance metrics that are commonly used across various sports to evaluate and quantify player and team performance. Here are some of the most common performance metrics in sports:

1. Scoring Metrics:
   - Points: Total points scored by an individual player or team. This metric is fundamental in many sports, including basketball, soccer, and American football.
   - Goals: The number of goals scored by a player or team. Goals are a primary scoring metric in sports like soccer, hockey, and lacrosse.
   - Runs: Runs scored in cricket or baseball. It represents the basic unit of scoring in these sports.

2. Shooting and Accuracy Metrics:
   - Field Goal Percentage (FG%): The percentage of successful field goals in basketball or soccer. It measures shooting accuracy.
   - Shooting Percentage: A similar metric used in sports like hockey and lacrosse to measure the accuracy of shots on goal.

3. Passing Metrics:
   - Assists: The number of assists made by a player in sports like basketball, soccer, and ice hockey. Assists measure the player's ability to create scoring opportunities for teammates.
   - Completion Percentage: Used in sports like American football and soccer to measure the accuracy of passes or throws completed.

4. Rebounding and Possession Metrics:
   - Rebounds: The number of rebounds grabbed by a player in basketball, soccer, or American football. Rebounds indicate a player's ability to secure possession.
   - Possession Time: The amount of time a team or player has control of the ball or puck. Possession time can be crucial in sports like soccer and ice hockey.

5. Defensive Metrics:
    - Tackles: The number of successful tackles made by a player in sports like football (both American and association), rugby, and Australian rules football.
    - Blocks: The number of blocked shots or attempts by a player in sports like basketball and soccer. It reflects defensive prowess.

6. Speed and Movement Metrics:
    - Sprint Speed: Measures how fast a player can sprint over a short distance, relevant in sports like soccer and American football.
    - Distance Covered: The total distance covered by a player during a game, often tracked using GPS technology. This metric is important in endurance sports like soccer and rugby.

7. Efficiency Metrics:
    - Player Efficiency Rating (PER): Used in basketball to measure a player's overall efficiency, taking into account various statistics like points, rebounds, assists, steals, and blocks.
    - Quarterback Rating (QBR): A metric used in American football to assess a quarterback's overall performance, considering passing accuracy, touchdowns, interceptions, and other factors.

8. Goalkeeper/Goalie Metrics:
    - Saves: The number of shots saved by a goalkeeper or goalie in sports like soccer and ice hockey.
    - Save Percentage: The percentage of shots saved by a goalkeeper relative to the total number of shots faced.

9. Win-Loss Record:
    - The number of games won and lost by a team or player. Win-loss records are essential for assessing overall success and competitiveness.

These are just a few examples of common performance metrics in sports. The specific metrics used can vary widely based on the sport, position, and the particular aspects of performance that are most relevant for evaluation. Sports analytics professionals often employ a combination of these metrics to gain a comprehensive understanding of player and team performance.

Creating custom metrics for player evaluation

Creating custom metrics for player evaluation in sports analytics allows teams and analysts to tailor performance assessments to their specific needs and strategies. Custom metrics can capture nuanced aspects of player performance that standard metrics may not fully address. Here's how you can create custom metrics for player evaluation:

1. Define the Objective:
   - Clearly define the specific objective or aspect of player performance you want to measure. For example, you might want to assess a basketball player's ability to create scoring opportunities for teammates beyond traditional assists.

2. Collect Relevant Data:
   - Gather relevant data points or statistics that are related to the objective you defined. This data could be game-specific, season-specific, or career-specific, depending on your goals.

3. Formulate the Metric:
   - Create a mathematical formula or algorithm that combines the collected data points into a single metric. This formula should reflect the desired aspect of player performance. Consider using weighted averages, ratios, or other mathematical operations to create the metric.

4. Normalize or Standardize (if necessary):
   - Depending on the metric's nature, you may need to normalize or standardize it to ensure it is on a consistent scale. This step can be essential when combining data with different units or scales.

5. Validate and Test:
   - Validate the custom metric by testing it against historical data or benchmarking it against existing performance metrics. Ensure that the metric aligns with your intended objective and provides meaningful insights.

6. Interpret and Visualize:
   - Interpret the results of your custom metric and create visualizations, such as charts or graphs, to make it more accessible and understandable to coaches, players, or team executives.

7. Iterate and Refine:
   - Continuously assess and refine your custom metric based on feedback, changes in game dynamics, or evolving player roles and strategies. Custom metrics should be adaptable over time.

Examples of Custom Metrics:
Here are a few examples of custom metrics for player evaluation:

Effective Ball Distribution (EBD) for Soccer:
- Formula: EBD = (Successful Passes into the Final Third) / (Total Passes Attempted)
- Objective: Assess a midfielder's ability to distribute the ball effectively into the final third of the field, contributing to scoring opportunities.

Shot Creation Efficiency (SCE) for Basketball:
- Formula: SCE = (Assists + Potential Assists) / (Field Goals Attempted)
- Objective: Evaluate a player's overall impact on creating scoring opportunities for teammates, accounting for both assists and potential assists.

Defensive Impact Rating (DIR) for American Football:
- Formula: DIR = (Tackles + Sacks + Interceptions) / (Opponent Yards Gained)
- Objective: Measure a defensive player's impact on preventing the opponent from gaining yards or scoring.

Pressure Index (PI) for Soccer:
- Formula: PI = (Total Pressures) / (Distance Covered)
- Objective: Assess a forward's or midfielder's ability to apply pressure on opponents by considering their work rate and coverage on the field.

Custom metrics should align with the strategic goals and priorities of the team or organization. They can be valuable tools for gaining a competitive edge and providing a more comprehensive view of player performance beyond traditional statistics.

Comparative analysis of players and teams

Comparative analysis of players and teams is a fundamental aspect of sports analytics, allowing teams, coaches, and analysts to evaluate and compare performance, make informed decisions, and develop winning strategies. Here's a step-by-step guide on how to conduct a comparative analysis of players and teams:
1. Define the Analysis Goals:
- Clearly define the objectives of your comparative analysis. Determine what specific aspects of player and team performance you want to evaluate and compare. Common objectives include assessing scoring ability, defensive prowess, teamwork, and overall efficiency.

2. Data Collection:
   - Gather relevant data for both players and teams. Depending on your goals, this data may include player statistics (e.g., points, rebounds, assists), team statistics (e.g., wins, goals scored), and any other performance-related metrics.

3. Data Cleaning and Preprocessing:
   - Clean and preprocess the data to ensure accuracy and consistency. Handle missing values, outliers, and data transformations as needed.

4. Define Performance Metrics:
   - Determine the performance metrics or indicators you will use to evaluate players and teams. These could be standard metrics (e.g., points per game, goals against average) or custom metrics tailored to your analysis goals.

5. Calculate Metrics:
   - Calculate the selected performance metrics for both players and teams. Ensure that the metrics are calculated consistently and accurately.

6. Visualization:
   - Create visualizations to compare players and teams. Use various types of charts and graphs to illustrate performance, such as bar charts, scatter plots, radar charts, or heatmaps.

7. Player vs. Player Comparison:
   - Compare individual player performance metrics to identify strengths and weaknesses relative to each other. For example, you might compare the scoring efficiency of two basketball players.

8. Team vs. Team Comparison:
   - Conduct comparative analysis at the team level to assess overall team performance. Compare team statistics, such as points scored, goals conceded, or win-loss records.

9. Player vs. Team Interaction:
   - Explore how individual player performance contributes to team success. Analyze how specific player metrics correlate with team performance metrics.

10. Identify Trends and Patterns: - Look for trends and patterns in the data. Are there players or teams that consistently outperform or underperform relative to others? Are there performance trends over time?

11. Statistical Analysis: - Perform statistical tests, such as t-tests, ANOVA, or correlation analysis, to assess the significance of differences or relationships between players and teams.

12. Machine Learning (if applicable): - Consider using machine learning models for predictive analysis. For example, you can build models to predict team wins based on player statistics.

13. Expert Insights: - Seek input from coaches, scouts, or sports experts to provide qualitative insights and context to the quantitative analysis.

14. Interpretation and Recommendations: - Interpret the results of your analysis and provide actionable recommendations. For example, you might recommend lineup changes, player rotations, or strategic adjustments based on the comparative analysis.

15. Visualization and Reporting: - Create clear and informative visual reports or dashboards to present your findings to stakeholders, including coaches, team management, and players.

16. Continuous Monitoring: - Continuously monitor player and team performance and update your analysis as new data becomes available. This allows for ongoing improvement and adjustment of strategies.

Comparative analysis of players and teams is an ongoing process in sports analytics. It can provide valuable insights for making decisions related to player development, roster management, game strategy, and overall team performance improvement.

Player evaluation models and case studies

Player evaluation models in sports analytics are designed to assess and quantify the performance, potential, and value of individual athletes. These models use a combination of statistical analysis, data-driven techniques, and sometimes machine learning to provide insights for teams, coaches, and front offices. Here are some common player evaluation models and case studies from various sports:

1. Player Efficiency Rating (PER) in Basketball:
   - Model Description: PER is a comprehensive player evaluation metric in basketball that takes into account various player statistics, such as points, rebounds, assists, steals, and blocks. It aims to provide a single number representing a player's overall contribution to the team.
   - Case Study: A case study could involve calculating PER for multiple NBA players over a season and comparing their ratings to assess their relative contributions to their respective teams.

2. Wins Above Replacement (WAR) in Baseball:
   - Model Description: WAR is a baseball player evaluation model that estimates how many additional wins a player contributes to their team compared to a replacement-level player. It considers offensive, defensive, and pitching contributions.
   - Case Study: Analyze the WAR values of different baseball players over a season to identify the most valuable players in terms of their impact on their teams' success.

3. Expected Goals (xG) in Soccer:
   - Model Description: xG is a soccer player evaluation model that estimates the likelihood of a shot resulting in a goal based on factors such as shot location, angle, and defensive pressure. It helps assess a player's ability to create scoring opportunities.
   - Case Study: Evaluate the xG values for individual soccer players to determine their effectiveness in creating high-quality scoring chances.

4. Quarterback Rating (QBR) in American Football:
   - Model Description: QBR is a comprehensive quarterback evaluation model in American football that considers passing accuracy, touchdown passes, interceptions, rushing ability, and situational performance. It provides a single rating for a quarterback's overall effectiveness.
   - Case Study: Assess the QBR of NFL quarterbacks over a season to rank their performance and identify key contributors to their teams' success.

5. Player Tracking Data Models in Basketball:
   - Model Description: Player tracking data models in basketball use advanced analytics and machine learning to analyze player movement, positioning, and performance on the court. These models can provide insights into player defensive coverage, offensive spacing, and shot selection.
   - Case Study: Analyze player tracking data to assess how a team's defensive strategy affects opponent shot percentages or to evaluate a player's impact on offensive spacing.

6. Pitcher Analysis in Cricket:
   - Model Description: In cricket, player evaluation models can assess a bowler's effectiveness based on factors like bowling speed, line and length, and variations. These models help teams make informed decisions about their bowling lineup.
   - Case Study: Evaluate the performance of fast bowlers in a cricket league by considering their wicket-taking ability, economy rate, and impact in different match situations.

7. Player Development Models:
   - Model Description: Player development models aim to assess a player's potential for improvement and future success. They consider factors like age, training, physical attributes, and historical performance.
   - Case Study: Analyze the development paths of young athletes in a sports academy to identify factors that contribute to their progression to elite levels.

These player evaluation models and case studies showcase the diverse range of approaches used in sports analytics to assess player performance, value, and potential. They provide valuable insights for talent scouting, roster management, game strategy, and overall team success.

Here's an example of player evaluation models and case studies in R, showcasing diverse approaches used in sports analytics to assess player performance. In this example, we'll focus on evaluating soccer (football) players, but similar approaches can be applied to other sports as well. We'll use a simplified dataset for demonstration purposes.

```r
# Load required libraries
library(tidyverse)
library(caret)
library(randomForest)

# Load soccer player performance data (example dataset)
soccer_data <- read.csv("soccer_player_data.csv")

# Data Preprocessing
# Assuming you have data cleaning, feature engineering, and player rating calculation steps

# Split the data into training and testing sets
set.seed(123)
train_indices <- createDataPartition(soccer_data$rating, p = 0.8, list = FALSE)
train_data <- soccer_data[train_indices, ]
test_data <- soccer_data[-train_indices, ]
# Player Evaluation Model (Example: Random Forest Regressor)
# Predict player ratings based on various performance metrics
rf_model <- randomForest(rating ~ goals + assists + interceptions + tackles + passing_accuracy,
            data = train_data, ntree = 500)

# Model Evaluation
rf_predictions <- predict(rf_model, newdata = test_data)
rmse <- sqrt(mean((rf_predictions - test_data$rating)^2))
```

```r
# Display the Root Mean Squared Error (RMSE) as a measure of model performance
cat("Random Forest Model RMSE:", rmse, "\n")

# Case Study: Identifying Underrated Players
# Analyze players who outperform their predicted ratings
underrated_players <- test_data %>%
  mutate(predicted_rating = rf_predictions) %>%
  filter(rating - predicted_rating > 5)  # Adjust the threshold as needed

# Display the list of underrated players
cat("Underrated Players:\n")
print(underrated_players)

# Case Study: Player Comparison
# Compare the performance of two players using the model
player1 <- data.frame(goals = 10, assists = 8, interceptions = 20, tackles = 30, passing_accuracy = 85)
player2 <- data.frame(goals = 15, assists = 6, interceptions = 18, tackles = 35, passing_accuracy = 90)

# Predict ratings for both players
player1_rating <- predict(rf_model, newdata = player1)
player2_rating <- predict(rf_model, newdata = player2)

# Display the predicted ratings for both players
cat("Player 1 Predicted Rating:", player1_rating, "\n")
cat("Player 2 Predicted Rating:", player2_rating, "\n")
```

In this example:

> We load soccer player performance data and preprocess it (data cleaning, feature engineering, and player rating calculation steps are assumed).
> We split the data into training and testing sets.
>
> We use a Random Forest Regressor model to predict player ratings based on performance metrics like goals, assists, interceptions, tackles, and passing accuracy.
>
> We evaluate the model's performance using Root Mean Squared Error (RMSE).
> We conduct two case studies: identifying underrated players and comparing the predicted ratings of two players based on their performance metrics.

These approaches can provide valuable insights for talent scouting, roster management, and game strategy in sports analytics by identifying players who outperform their ratings and comparing the potential of different players.

Chapter 5: Team Analysis and Strategy

Team-level performance metrics

Team-level performance metrics are essential in sports analytics to assess and evaluate the overall effectiveness, success, and efficiency of sports teams. These metrics help teams, coaches, and analysts understand team dynamics, identify strengths and weaknesses, and make data-driven decisions. Here are some common team-level performance metrics used in various sports:

1. Win-Loss Record:
   - The simplest and most fundamental team performance metric. It tracks the number of games won and lost by the team over a specific period, such as a season.

2. Winning Percentage:
   - Calculated as (Wins / Total Games Played). It provides a normalized measure of a team's success, allowing for easier comparisons between teams with different numbers of games played.

3. Points Scored:
   - The total number of points or goals scored by the team in a season or game. This metric is crucial in sports like basketball, soccer, and hockey.

4. Points Allowed (Goals Conceded):
   - The total number of points or goals allowed by the team's defense or goalkeeper. It measures defensive performance and is vital in sports like soccer and American football.

5. Scoring Efficiency:
   - Calculated as (Points Scored / Total Possessions or Shots Taken). It quantifies how efficiently a team converts scoring opportunities into points.

6. Goal Differential:
   - Calculated as (Goals Scored - Goals Conceded). A positive goal differential indicates that a team scores more than it concedes, while a negative differential suggests the opposite.

7. Average Points Per Game (PPG):
   - The average number of points a team scores per game in a season. Commonly used in basketball and other high-scoring sports.

8. Average Goals Against (GA):
   - The average number of goals conceded per game in a season, indicating defensive strength or vulnerability.

9. Possession Percentage:
   - The percentage of time a team has control of the ball or puck during a game. It can indicate a team's ability to maintain possession and control the game's tempo.

10. Passing Accuracy: - The percentage of successful passes completed by the team. It reflects the team's ability to maintain ball movement and control.

11. Offensive Efficiency Rating: - A composite metric that considers points scored, field goal percentage, turnovers, and other offensive statistics to assess a team's offensive performance.

12. Defensive Efficiency Rating: - A composite metric that considers points allowed, opponent field goal percentage, steals, and other defensive statistics to assess a team's defensive performance.

13. Rebounding Percentage: - The percentage of available rebounds grabbed by a team. It reflects a team's ability to control the boards in sports like basketball and soccer.

14. Turnover Differential: - Calculated as (Forced Turnovers - Turnovers Committed). It measures a team's ability to force turnovers while limiting their own.

15. Red Zone Efficiency (RZE): - In American football, RZE measures a team's success in scoring touchdowns when in the opponent's red zone (inside the 20-yard line).

16. Penalty Kill Percentage (PK%): - In ice hockey, PK% measures a team's success in preventing goals when shorthanded due to penalties.

17. Power Play Percentage (PP%): - In ice hockey, PP% measures a team's success in scoring goals when having a player advantage due to an opponent's penalty.

These team-level performance metrics provide a comprehensive view of a team's performance in various aspects of the game. Sports analysts often use a combination of these metrics to conduct in-depth assessments, identify areas for improvement, and make strategic decisions to enhance team performance.

Analyzing team dynamics and strategy

Analyzing team dynamics and strategy in sports involves a multifaceted approach that combines data analysis, observation, and expert insights. Teams, coaches, and analysts use various methods to assess how a team operates on and off the field and develop strategies for improvement. Here's a step-by-step guide on how to analyze team dynamics and strategy:

1. Data Collection:
    - Gather relevant data sources, including game statistics, player performance metrics, video footage, and tracking data (if available). These sources provide the foundation for analysis.

2. Define Objectives:
    - Clearly define the objectives of your analysis. What aspects of team dynamics and strategy do you want to assess? Common objectives include offensive and defensive strategies, player roles, and teamwork.

3. Performance Metrics:
    - Select performance metrics and indicators that align with your objectives. These may include scoring efficiency, possession time, passing accuracy, shot selection, defensive pressure, and more.

4. Video Analysis:
    - Review game footage to assess player movements, tactics, and decision-making. Identify patterns in how the team executes offensive and defensive plays.

5. Statistical Analysis:
    - Perform statistical analysis to assess team performance. This may involve calculating averages, percentages, correlations, and advanced metrics to uncover insights.

6. Benchmarking:
    - Compare your team's performance metrics to those of other teams in the same league or similar competitions. Benchmarking can reveal strengths and weaknesses relative to competitors.

7. Identify Patterns and Trends:
   - Look for recurring patterns and trends in team behavior. For example, analyze whether the team tends to score more goals in the second half of games or if there are specific formations that yield better results.

8. Player Roles and Contributions:
   - Assess the roles and contributions of individual players within the team. Identify key playmakers, defenders, and those who excel in specific situations.

9. Tactical Analysis:
   - Analyze the team's tactical approach, such as offensive formations, defensive strategies, and set-piece plays. Evaluate how well the team executes these tactics.

10. Game Situation Analysis:
- Study how the team performs in different game situations, such as when leading, trailing, or tied. Assess the team's adaptability and decision-making under pressure.

11. Heatmaps and Spatial Analysis:
- Use heatmaps and spatial analysis to visualize player and ball movement on the field. This can reveal areas of dominance and identify where the team needs improvement.

12. Player-Team Dynamics:
- Examine how players interact and collaborate on the field. Assess the effectiveness of player-to-player and player-to-coach communication.

13. Expert Input:
- Consult with coaches, sports psychologists, and other experts who can provide qualitative insights into team dynamics, motivation, and mental aspects of the game.

14. Strategy Development:
- Based on your analysis, work with the coaching staff to develop strategies and tactics that capitalize on strengths and address weaknesses. Consider both short-term and long-term strategies.

15. Player Development and Training:
- Tailor training programs to address specific weaknesses or areas of improvement identified in the analysis. Provide feedback and guidance to players.

16. Continuous Monitoring:
- Continuously monitor team dynamics and strategy throughout the season. Adjust strategies as needed based on performance and opponent analysis.

17. Visualization and Reporting:
- Create visual reports and presentations to communicate your findings and recommendations to team stakeholders, including coaches, players, and management.

Analyzing team dynamics and strategy is an ongoing process that evolves with the team's development and the changing dynamics of the sport. Data-driven insights and expert input combine to inform decision-making, enhance performance, and contribute to a winning strategy.

Analyzing team dynamics and strategy in sports analytics often involves assessing how a team performs collectively and how various factors impact team success. Below, I'll provide an example in R using a simplified dataset for analyzing soccer (football) team dynamics and strategy.

```r
# Load required libraries
library(tidyverse)

# Load soccer team performance data (example dataset)
team_data <- read.csv("soccer_team_data.csv")

# Data Preprocessing
# Assuming you have data cleaning and feature engineering steps

# Analyzing Team Dynamics
# Calculate team-level metrics and trends
team_metrics <- team_data %>%
  group_by(team_id) %>%
  summarise(
    avg_goals_scored = mean(goals_scored),
    avg_goals_conceded = mean(goals_conceded),
    avg_possession = mean(possession),
    avg_pass_completion = mean(pass_completion)
  )

# Visualizing Team Dynamics
# Create visualizations to understand team performance over time
ggplot(team_metrics, aes(x = team_id)) +
  geom_line(aes(y = avg_goals_scored), color = "blue") +
  geom_line(aes(y = avg_goals_conceded), color = "red") +
  labs(title = "Average Goals Scored and Conceded Over Time",
       x = "Team ID",
       y = "Average Goals")
```

```r
ggplot(team_metrics, aes(x = team_id)) +
  geom_line(aes(y = avg_possession), color = "green") +
  geom_line(aes(y = avg_pass_completion), color = "purple") +
  labs(title = "Average Possession and Pass Completion Over Time",
     x = "Team ID",
     y = "Average Percentage")

# Analyzing Team Strategy
# Identify common strategies based on performance metrics
strategy_clusters <- kmeans(team_metrics[, -1], centers = 3, nstart = 20)

# Add cluster labels to the team_metrics data
team_metrics$strategy_cluster <- factor(strategy_clusters$cluster)

# Visualize team strategy clusters
ggplot(team_metrics, aes(x = avg_goals_scored, y = avg_goals_conceded, color = strategy_cluster)) +
  geom_point() +
  labs(title = "Team Strategy Clusters",
     x = "Average Goals Scored",
     y = "Average Goals Conceded")

# Interpret the clusters and their strategies

# Team Strategy Recommendations
# Based on cluster analysis, provide recommendations for team strategy adjustments
cluster1_teams <- team_metrics %>%
  filter(strategy_cluster == 1)

cluster2_teams <- team_metrics %>%
  filter(strategy_cluster == 2)

cluster3_teams <- team_metrics %>%
  filter(strategy_cluster == 3)

# Display strategy recommendations for each cluster

# Additional analyses can include player-level impact on team dynamics, tactical assessments, and more.
```

In this example:

> We load soccer team performance data and preprocess it (data cleaning and feature engineering are assumed).
>
> We calculate team-level metrics such as average goals scored, average goals conceded, average possession, and average pass completion.
>
> We create visualizations to understand team performance trends over time.
>
> We apply clustering (e.g., k-means clustering) to identify common team strategies based on performance metrics.
>
> We interpret the clusters and provide recommendations for team strategy adjustments.

Analyzing team dynamics and strategy in this way can provide insights into how teams perform collectively and help teams make data-driven decisions to optimize their strategies for success.

Game simulation and strategy optimization

Game simulation and strategy optimization are powerful tools in sports analytics that allow teams and coaches to simulate game scenarios, test different strategies, and make data-driven decisions to improve performance. Here's a step-by-step guide on how to use game simulation and strategy optimization in sports:

1. Data Collection and Analysis:
   - Collect historical game data, player performance metrics, and other relevant statistics. Analyze this data to understand team strengths, weaknesses, and historical performance patterns.

2. Define Simulation Objectives:
   - Clearly define the objectives of your game simulation. What specific scenarios or strategies do you want to test or optimize? Examples include lineup changes, offensive plays, defensive formations, and in-game decisions.

3. Develop a Simulation Model:
   - Create a simulation model that replicates the game dynamics, including player interactions, ball movement, and scoring. The model should be based on historical data and adjusted for specific scenarios.

4. Input Parameters:
   - Define the input parameters for the simulation. These parameters might include player attributes, opponent strengths, field conditions, and game situations (e.g., score, time remaining).

5. Simulate Game Scenarios:
   - Run the simulations to simulate various game scenarios. This can involve thousands of simulations to explore a wide range of possibilities.

6. Evaluate Performance Metrics:
   - Measure the outcomes and performance metrics of each simulated scenario. Common metrics include win probability, expected points, and scoring efficiency.

7. Strategy Optimization:
   - Use optimization algorithms to find the best strategies or decisions that maximize desired outcomes. Optimization may involve maximizing scoring, minimizing opponent points, or achieving specific game objectives.

8. Sensitivity Analysis:
   - Perform sensitivity analysis to understand how changes in input parameters impact the results. This helps assess the robustness of optimized strategies.

9. Visualization and Reporting:
   - Visualize the results of the simulations and strategy optimization. Create reports and dashboards to present the findings to coaches and team stakeholders.

10. Strategy Implementation:
- Translate the optimized strategies into actionable game plans. Work with coaches and players to ensure they understand and can execute the chosen strategies.

11. In-Game Decision Support:
- During actual games, use real-time data and the insights gained from simulations to inform in-game decisions. This may involve substitutions, play calling, and tactical adjustments.

12. Continuous Improvement:
- Continuously update and refine the simulation model and optimization strategies as new data becomes available and as the team's dynamics evolve.

Examples of Game Simulation and Strategy Optimization:

Basketball:
- Simulate different offensive plays and strategies to find the most effective ways to score against different opponents' defenses.

Soccer:
- Optimize player positioning and passing strategies to maximize goal-scoring opportunities while minimizing defensive vulnerabilities.

American Football:
- Simulate different play-calling scenarios to optimize offensive and defensive strategies based on game situations.

Baseball:
- Use simulations to optimize pitching rotations and batting lineups for the best chance of winning games.

Hockey:
- Simulate power plays, penalty kills, and line combinations to optimize scoring efficiency and defensive performance.

Game simulation and strategy optimization in sports analytics provide a systematic and data-driven approach to improving team performance. By testing strategies in a risk-free environment and optimizing decision-making, teams can gain a competitive edge and increase their chances of success on the field or court.

Game simulation and strategy optimization in sports analytics involve using models and simulations to make data-driven decisions and find optimal strategies. Here's an example in R using a simplified basketball simulation for strategy optimization:

```r
# Load required libraries
library(dplyr)
library(ggplot2)

# Set the simulation parameters
num_simulations <- 1000  # Number of game simulations
team_a_strength <- 85    # Team A's strength (scale of 0 to 100)
team_b_strength <- 80    # Team B's strength (scale of 0 to 100)

# Initialize variables to track game outcomes
team_a_wins <- 0
team_b_wins <- 0
draws <- 0
```

```r
# Simulate basketball games
for (i in 1:num_simulations) {
  # Simulate game outcome using a simple strength-based model
  team_a_score <- rpois(1, team_a_strength / 10)
  team_b_score <- rpois(1, team_b_strength / 10)

  # Determine the winner or declare a draw
  if (team_a_score > team_b_score) {
    team_a_wins <- team_a_wins + 1
  } else if (team_b_score > team_a_score) {
    team_b_wins <- team_b_wins + 1
  } else {
    draws <- draws + 1
  }
}

# Calculate win percentages
team_a_win_percentage <- team_a_wins / num_simulations * 100
team_b_win_percentage <- team_b_wins / num_simulations * 100
draw_percentage <- draws / num_simulations * 100

# Print the simulation results
cat("Simulation Results (", num_simulations, " games simulated):\n")
cat("Team A Win Percentage: ", team_a_win_percentage, "%\n")
cat("Team B Win Percentage: ", team_b_win_percentage, "%\n")
cat("Draw Percentage: ", draw_percentage, "%\n")

# Visualize the results
results_df <- data.frame(Team = c("Team A", "Team B", "Draw"),
                         Percentage = c(team_a_win_percentage,
team_b_win_percentage, draw_percentage))

ggplot(results_df, aes(x = Team, y = Percentage, fill = Team)) +
  geom_bar(stat = "identity") +
  labs(title = "Simulation Results",
       y = "Percentage",
       fill = "Team")
```

In this basketball simulation example:

> We set the simulation parameters, including the number of simulations and the strength of two basketball teams (Team A and Team B).

We simulate game outcomes using a simple strength-based model. The rpois function generates random scores based on team strength.

We track the number of wins for each team and draws in the simulations.

We calculate win percentages for Team A and Team B, as well as the percentage of draws.

We print the simulation results and visualize them using a bar chart.

This is a basic example, and in real-world sports analytics, simulations can be much more complex, involving factors like player performance, tactical decisions, and more advanced models. The goal is to use simulations to optimize strategies and make informed decisions in sports.

Case studies of successful team strategies

Successful team strategies in sports can vary widely depending on the sport, the team's strengths, and the specific circumstances of a game or season. Here are some case studies of successful team strategies in different sports:

1. "Moneyball" Strategy in Baseball:
   - Sport: Baseball
   - Case Study: The Oakland Athletics' "Moneyball" strategy, popularized by the book and movie of the same name, focused on using advanced statistics to identify undervalued players and build a competitive team on a limited budget. The team's emphasis on on-base percentage and walks led to success despite having a lower payroll.

2. "Total Football" by Ajax and Barcelona:
   - Sport: Soccer (Football)
   - Case Study: Teams like Ajax and Barcelona have implemented "Total Football" strategies that prioritize ball possession, fluid passing, and positional play. This approach emphasizes player versatility, quick ball circulation, and a high-pressing game to dominate opponents.

3. "Triangle Offense" by Chicago Bulls:
   - Sport: Basketball
   - Case Study: The Chicago Bulls, led by Michael Jordan and coached by Phil Jackson, popularized the "Triangle Offense." This strategy maximized ball movement, spacing, and player positioning, allowing the Bulls to win multiple NBA championships in the 1990s.

4. "Moneyball" Strategy in Basketball (Houston Rockets):
   - Sport: Basketball
   - Case Study: The Houston Rockets have embraced a data-driven approach to basketball similar to the "Moneyball" philosophy in baseball. They heavily rely on three-point shooting and advanced analytics to optimize shot selection and player efficiency.

5. "Gegenpressing" in Soccer (Liverpool FC):
   - Sport: Soccer (Football)
   - Case Study: Liverpool FC, under the management of Jurgen Klopp, implemented a high-intensity pressing style known as "Gegenpressing." This strategy aims to win back possession quickly after losing it and counterattack with speed. It has led to domestic and international success for the club.

6. Defensive Dominance by the Baltimore Ravens:
   - Sport: American Football
   - Case Study: The Baltimore Ravens have a history of fielding dominant defensive teams. Their strategy focuses on creating turnovers, strong pass rushes, and suffocating pass coverage. This strategy has led to Super Bowl victories and consistent playoff appearances.

7. "Moneyball" Strategy in Hockey (Tampa Bay Lightning):
   - Sport: Ice Hockey
   - Case Study: The Tampa Bay Lightning have embraced analytics to build a competitive team. They prioritize puck possession and shot quality, using data to make roster decisions and develop game strategies. This approach has resulted in Stanley Cup victories.

8. "Inside-Out" Basketball Strategy (San Antonio Spurs):
   - Sport: Basketball
   - Case Study: The San Antonio Spurs are known for their "inside-out" basketball strategy. They focus on ball movement, post play, and three-point shooting. This approach, combined with strong team defense, has led to multiple NBA championships.

9. Tactical Flexibility by New England Patriots:
   - Sport: American Football
   - Case Study: The New England Patriots, led by coach Bill Belichick, are known for their adaptability and tactical flexibility. They adjust their game plans to exploit opponent weaknesses, making them a perennial contender in the NFL.

These case studies illustrate that successful team strategies can vary widely, but they often share common elements such as data-driven decision-making, adaptability, and a focus on maximizing the strengths of the team's personnel. The best strategies are tailored to the team's unique circumstances and are continuously refined based on performance and opponent analysis.

Chapter 6: Injury Analysis and Prevention

Injury data collection and analysis

Injury data collection and analysis are critical components of sports analytics that help teams and organizations understand the patterns, causes, and implications of injuries among athletes. Analyzing injury data allows teams to develop injury prevention strategies, optimize player health and performance, and make informed decisions. Here's how injury data collection and analysis work:

1. Data Collection:
   - Collecting comprehensive injury data is the first step. This includes documenting all injuries sustained by athletes, whether in practices or games. Key data points to collect include:
     - Player Information: Name, age, position, and injury history.
     - Injury Details: Type, location, severity, date, and mechanism of injury.
     - Context: Was the injury sustained during practice, a game, or off the field? Was it due to contact or non-contact?
     - Rehabilitation Data: Information on treatment, recovery time, and return-to-play protocols.

2. Injury Classification:
   - Categorize injuries by type (e.g., muscle strains, ligament injuries, fractures) and severity (e.g., minor, moderate, severe) to identify common patterns.

3. Data Management:
   - Maintain a centralized and secure database for injury data, ensuring that it is up-to-date, easily accessible, and organized for analysis.

4. Analysis Techniques:
   - Use statistical analysis and data visualization to gain insights from injury data. Common analysis techniques include:
     - Frequency Analysis: Determine the most common types of injuries and their occurrence rates.
     - Time Trends: Analyze when injuries are most likely to occur during a season.
     - Comparative Analysis: Compare injury rates between different teams, seasons, or player groups.
     - Risk Factors: Identify factors contributing to injuries, such as player workload, playing surface, and training methods.

5. Injury Risk Assessment:
   - Develop models or algorithms to assess injury risk for individual players. These models may consider player characteristics, training loads, and previous injury history.

6. Injury Prevention Strategies:
   - Based on the analysis, implement injury prevention strategies tailored to the specific risks identified. This may involve adjusting training regimens, modifying playing techniques, or improving equipment.

7. Monitoring and Reporting:
   - Continuously monitor injury data throughout the season and provide regular reports to coaches, medical staff, and team management. Highlight trends and patterns to inform decision-making.

8. Return-to-Play Protocols:
   - Develop evidence-based return-to-play protocols that consider injury type and severity. Ensure that players are fully recovered and have reduced injury risk before returning to action.

9. Player Load Management:
   - Manage player workloads to prevent overuse injuries. Monitoring training volume, intensity, and recovery is crucial to avoid physical stress that can lead to injuries.

10. Feedback Loops:
- Establish feedback loops with medical professionals, coaches, and players to continuously improve injury prevention strategies based on real-world data and experiences.

11. Research and Innovation:
- Collaborate with sports scientists, researchers, and medical experts to stay up-to-date with the latest research on injury prevention techniques and technologies.

12. Compliance and Ethics:
- Ensure that injury data collection and analysis comply with privacy regulations and ethical standards. Protect player confidentiality and sensitive medical information.

Injury data collection and analysis play a pivotal role in reducing the risk of injuries, enhancing player safety, and optimizing team performance. Teams and organizations that invest in comprehensive injury analytics can minimize downtime, maximize player availability, and maintain a competitive edge in sports.

Collecting and analyzing injury data is crucial in sports analytics to understand and mitigate injury risks for athletes. In this example, we'll demonstrate how to collect and analyze injury data using R. We'll use a hypothetical dataset of injury records.

```r
# Load required libraries
library(tidyverse)

# Load injury data (example dataset)
injury_data <- read.csv("injury_data.csv")

# Data Preprocessing
# Assuming you have data cleaning and feature engineering steps

# Explore the injury data
summary(injury_data)

# Calculate injury rates and trends
injury_rates <- injury_data %>%
  group_by(year, sport) %>%
  summarize(total_injuries = n()) %>%
  mutate(injury_rate = total_injuries / sum(total_injuries) * 100)

# Visualize injury rates over the years
ggplot(injury_rates, aes(x = year, y = injury_rate, color = sport)) +
  geom_line() +
  labs(title = "Injury Rates Over the Years",
       x = "Year",
       y = "Injury Rate (%)")

# Analyze injury types
injury_type_counts <- injury_data %>%
  group_by(injury_type) %>%
  summarize(count = n()) %>%
  arrange(desc(count))

# Visualize injury types
ggplot(injury_type_counts, aes(x = reorder(injury_type, -count), y = count)) +
  geom_bar(stat = "identity", fill = "blue") +
  theme(axis.text.x = element_text(angle = 45, hjust = 1)) +
  labs(title = "Injury Types",
       x = "Injury Type",
       y = "Count")

# Analyze injury severity
```

```
injury_severity_stats <- injury_data %>%
  group_by(injury_severity) %>%
  summarize(avg_recovery_time = mean(recovery_time_days, na.rm = TRUE))

# Visualize injury severity
ggplot(injury_severity_stats, aes(x = injury_severity, y = avg_recovery_time)) +
  geom_bar(stat = "identity", fill = "green") +
  labs(title = "Injury Severity and Average Recovery Time",
       x = "Injury Severity",
       y = "Average Recovery Time (days)")

# Identify high-risk periods or situations for injuries

# Provide recommendations for injury prevention and player safety
```

In this example:

> We load hypothetical injury data and preprocess it (data cleaning and feature engineering are assumed).

> We explore the injury data, including summary statistics, to understand the dataset.

> We calculate injury rates over the years for different sports and visualize them.

> We analyze injury types and visualize the distribution of injury types.

> We analyze injury severity and visualize the relationship between injury severity and average recovery time.

> We can further identify high-risk periods or situations for injuries and provide recommendations for injury prevention and player safety based on the analysis.

In real-world sports analytics, injury data analysis can be more extensive, involving advanced statistical modeling, machine learning for injury prediction, and collaboration with medical professionals to improve player safety.

Identifying injury risk factors

Identifying injury risk factors is crucial in sports to prevent injuries, optimize player health, and enhance team performance. Injuries can result from a combination of intrinsic (player-related) and extrinsic (environmental and situational) factors. Here are key steps and common risk factors to consider when identifying injury risk factors:

1. Data Collection:
   - Collect comprehensive injury data as described earlier, including player information, injury details, and contextual information.

2. Literature Review:
   - Review scientific literature and studies related to sports injuries in your specific sport. This can provide insights into known risk factors.

3. Statistical Analysis:
   - Analyze the injury data using statistical methods to identify potential risk factors. Consider the following types of risk factors:

Intrinsic Risk Factors (Player-Related):
   - Age: Younger and older athletes may be more susceptible to certain injuries.
   - Previous Injuries: Athletes with a history of similar injuries are at increased risk.
   - Body Composition: Factors such as body mass index (BMI) and muscle imbalances can contribute to injury risk.
   - Biomechanics: Evaluate player movement patterns and biomechanics to identify abnormal mechanics that increase injury risk.
   - Flexibility and Range of Motion: Reduced flexibility or limited range of motion can increase the risk of certain injuries.
   - Strength and Conditioning: Weaknesses or imbalances in muscle groups can lead to injuries.
   - Training Load: Excessive or insufficient training loads and volumes can increase the risk of overuse or acute injuries.
   - Fitness Level: Players with lower fitness levels may be at greater risk.
   - Position: Certain positions within a sport may be associated with higher injury risk due to different physical demands.
   - Psychological Factors: Mental stress, anxiety, and motivation can influence injury risk.

Extrinsic Risk Factors (Environmental and Situational):
   - Playing Surface: The type and condition of the playing surface can affect injury risk.

- Equipment: Poorly fitting or outdated equipment can contribute to injuries.
- Weather Conditions: Extreme weather conditions, such as heat, cold, or rain, can increase injury risk.
- Opponent Behavior: Aggressive opponents or fouls may increase the risk of contact injuries.
- Game Situation: The intensity and importance of a game or competition may affect player decision-making and injury risk.

4. Multifactorial Analysis:
- Recognize that injuries often result from the interaction of multiple risk factors. Consider the interplay between intrinsic and extrinsic factors.

5. Machine Learning and Predictive Modeling:
- Develop predictive models using machine learning techniques to assess injury risk. These models can incorporate multiple risk factors and provide a quantitative assessment of risk.

6. Medical Evaluation:
- Include medical assessments, such as physical examinations and imaging, to identify underlying physical conditions or vulnerabilities that may increase injury risk.

7. Regular Monitoring:
- Continuously monitor players' health, workload, and injury history throughout the season. Adjust training and recovery strategies based on this information.

8. Collaboration:
- Collaborate with sports scientists, medical professionals, and coaching staff to collectively identify and address injury risk factors.

9. Injury Prevention Programs:
- Develop injury prevention programs that specifically target identified risk factors. These programs may include strength and conditioning, biomechanical training, and sports psychology interventions.

10. Education:
- Educate players and coaches about the identified risk factors and the importance of injury prevention strategies.

Identifying injury risk factors is an ongoing process that involves a combination of data analysis, scientific research, and practical insights. By identifying and addressing these risk factors, sports organizations can reduce the likelihood of injuries, improve player longevity, and maintain a competitive edge.

Injury prediction models

Injury prediction models in sports aim to forecast the likelihood of a player sustaining an injury based on various risk factors and historical data. These models are valuable tools for sports teams, medical staff, and coaches to proactively manage player health and reduce injury risk. Here's an overview of how injury prediction models work and factors to consider when developing them:

1. Data Collection:
   - Gather comprehensive injury data, including player information, injury history, injury types, and contextual factors such as training load and playing conditions. Historical data is crucial for training and evaluating the prediction model.

2. Feature Selection:
   - Identify relevant features (variables) that may contribute to injury risk. These features can include both intrinsic (player-related) and extrinsic (environmental and situational) factors. Common features include player age, previous injuries, workload, biomechanics, and psychological factors.

3. Data Preprocessing:
   - Clean and preprocess the data to handle missing values, outliers, and standardize the features. Proper data preprocessing ensures that the model performs accurately.

4. Model Selection:
   - Choose an appropriate machine learning or statistical model for injury prediction. Common models used for injury prediction include logistic regression, decision trees, random forests, support vector machines, and neural networks. The choice of model may depend on the complexity of the problem and the available data.

5. Training the Model:
   - Split the dataset into a training set and a validation set. Train the model on the training set and tune hyperparameters to optimize performance. Use the validation set to assess the model's generalization ability.

6. Model Evaluation:
   - Assess the model's performance using relevant evaluation metrics such as accuracy, precision, recall, F1-score, and area under the receiver operating characteristic curve (AUC-ROC). These metrics provide insights into the model's ability to correctly predict injuries and minimize false alarms.

7. Cross-Validation:
   - Perform cross-validation to ensure that the model's performance is consistent across different subsets of the data. This helps mitigate overfitting.

8. Incorporating Time:
   - Consider temporal aspects of injuries by incorporating time-based features, such as the number of days since the last injury or the cumulative workload over a specific period.

9. Feature Importance Analysis:
   - Analyze feature importance to understand which risk factors have the most significant impact on injury prediction. This insight can inform injury prevention strategies.

10. Threshold Selection:
- Determine an appropriate threshold for classifying injury risk. Depending on the application, you may prioritize sensitivity (identifying as many at-risk players as possible) or specificity (minimizing false alarms).

11. Deployment:
- Integrate the trained injury prediction model into the team's workflow, allowing for real-time or periodic risk assessments of players.

12. Continuous Updating:
- Continuously update the model with new injury data and refine it based on evolving risk factors and trends.

13. Ethical Considerations:
- Address ethical concerns related to player privacy, informed consent, and the responsible use of injury prediction models.

14. Communication:
- Effectively communicate the results of injury predictions to coaches, medical staff, and players, emphasizing the importance of injury prevention strategies and player well-being.

Injury prediction models in sports are valuable tools for enhancing player health and performance. While they cannot guarantee injury prevention, they provide valuable insights that enable teams to take proactive measures to reduce the risk of injuries and ensure the well-being of their athletes.

Injury prevention strategies in sports

Injury prevention strategies in sports are essential to minimize the risk of injuries among athletes and maintain their long-term health and performance. These strategies involve a combination of measures focused on conditioning, training, monitoring, and education. Here are some key injury prevention strategies in sports:

1. Pre-Season Physical Examinations:
   - Conduct comprehensive pre-season medical assessments to identify any underlying health issues or physical vulnerabilities that could increase injury risk. This includes screening for cardiovascular conditions, musculoskeletal problems, and any previous injuries.

2. Proper Warm-Up and Cool-Down:
   - Implement a structured warm-up and cool-down routine before and after training sessions and games. Dynamic stretching, mobility exercises, and light aerobic activities can help prepare the body and reduce the risk of muscle strains and soft tissue injuries.

3. Strength and Conditioning Programs:
   - Develop individualized strength and conditioning programs that target specific muscle groups and address any muscle imbalances or weaknesses. A well-rounded program can improve overall fitness, stability, and injury resistance.

4. Periodization:
   - Employ periodization in training plans to vary the intensity and workload throughout the season. This helps prevent overuse injuries and ensures proper recovery.

5. Monitoring Workload:
   - Use workload monitoring systems to track athletes' training loads and recovery. Avoid rapid increases in workload that can lead to overtraining and increased injury risk.

6. Sports-Specific Training:
   - Incorporate sports-specific drills and exercises that mimic game situations and movements. This helps athletes adapt to the demands of their sport and reduce the risk of injuries related to unfamiliar movements.

7. Biomechanical Analysis:
   - Use biomechanical analysis to assess and correct faulty movement patterns that may contribute to injuries. This can include gait analysis, video analysis, and motion capture technology.

8. Injury Prevention Exercises:
   - Include injury prevention exercises in training programs. These exercises can target areas prone to injury, such as the knees, ankles, and shoulders. For example, neuromuscular training programs like the FIFA 11+ for soccer can reduce the risk of ACL injuries.

9. Sport-Specific Equipment:
   - Ensure that athletes are equipped with appropriate and well-fitting gear, including helmets, pads, footwear, and protective equipment. Regularly inspect and maintain equipment for safety.

10. Hydration and Nutrition: - Emphasize proper hydration and nutrition to support optimal performance and recovery. Dehydration and nutrient deficiencies can increase the risk of cramps and soft tissue injuries.

11. Sleep and Recovery: - Prioritize adequate sleep and recovery. Sleep is essential for tissue repair, muscle growth, and overall well-being. Encourage rest days and recovery activities like foam rolling and stretching.

12. Psychological Support: - Provide access to sports psychologists or mental health professionals to help athletes manage stress, anxiety, and performance-related pressure. Mental well-being is closely linked to injury prevention.

13. Environmental Considerations: - Consider environmental factors, such as weather conditions and playing surfaces. Adapt training and game strategies accordingly to reduce the risk of weather-related or surface-related injuries.

14. Education and Communication: - Educate athletes, coaches, and medical staff about injury prevention strategies and the importance of reporting injuries promptly. Encourage open communication and a culture of safety.

15. Injury Tracking and Analysis: - Continuously monitor and analyze injury data to identify trends and patterns. Use this information to refine injury prevention strategies and make data-driven decisions.

16. Athlete Education: - Educate athletes about proper technique, safe play, and the risks associated with certain movements or actions in their sport. Encourage responsible and sportsmanlike behavior.

Injury prevention is an ongoing process that requires a holistic approach, including physical preparation, education, and monitoring. By implementing these strategies, sports organizations can reduce the risk of injuries, enhance player performance, and promote the long-term health and well-being of athletes.

Chapter 7: Predictive Modeling in Sports

Introduction to predictive modeling

Predictive modeling is a powerful analytical technique used across various fields, including data science, business, finance, healthcare, and sports analytics. It involves the use of mathematical and statistical models to predict future outcomes or trends based on historical data and patterns. Predictive modeling aims to make informed forecasts and decisions by leveraging data-driven insights. Here's an introduction to predictive modeling:

Key Concepts:

Historical Data: Predictive modeling starts with historical data, which serves as the foundation for building and training the predictive model. This data includes past observations, measurements, or events relevant to the problem being addressed.

Predictive Variables (Features): These are the variables or factors from the historical data that are used as inputs to the predictive model. These variables can be quantitative (e.g., numerical values) or categorical (e.g., types or categories).

Target Variable: The target variable is the specific outcome or event that the predictive model aims to predict. It's the variable of interest, and its values can be continuous (e.g., predicting sales revenue) or binary (e.g., predicting customer churn).

Model Building: Predictive modeling involves selecting an appropriate mathematical or statistical model based on the characteristics of the data and the problem. Common models include linear regression, decision trees, random forests, support vector machines, neural networks, and more.

Training the Model: The predictive model is trained using the historical data. During training, the model learns the relationships and patterns between the predictive variables and the target variable.

Validation and Testing: After training, the model is validated and tested on new, unseen data to assess its performance. This step helps ensure that the model can generalize its predictions to new situations effectively.

Performance Metrics: To evaluate the model's performance, various metrics are used, depending on the nature of the problem. Common metrics include accuracy, precision, recall, F1-score, mean squared error (MSE), and area under the receiver operating characteristic curve (AUC-ROC).

Applications:
Predictive modeling has a wide range of applications, including:

- Financial Forecasting: Predicting stock prices, credit risk assessment, and fraud detection.
- Healthcare: Predicting disease outcomes, patient readmission risk, and treatment responses.
- Marketing: Targeted advertising, customer churn prediction, and recommendation systems.
- Manufacturing: Predictive maintenance to avoid equipment breakdowns.
- Sports Analytics: Predicting game outcomes, player performance, and injury risk.
- Environmental Science: Predicting weather patterns and natural disasters.
- E-commerce: Demand forecasting and inventory optimization.

Challenges:
Predictive modeling comes with several challenges, including:

- Data Quality: The quality of historical data is crucial. Inaccurate or incomplete data can lead to unreliable predictions.
- Overfitting: A model that is overly complex can fit the training data perfectly but perform poorly on new data due to overfitting.
- Bias: Biased data or biased model assumptions can lead to unfair or discriminatory predictions.
- Data Imbalance: Imbalanced datasets can lead to poor model performance, especially when predicting rare events.

Ethical Considerations:
Predictive modeling can have ethical implications, especially in areas like healthcare, finance, and criminal justice. Ensuring fairness, transparency, and accountability in modeling practices is essential to avoid biased or discriminatory outcomes.

In summary, predictive modeling is a data-driven approach used to make predictions about future events or outcomes based on historical data and patterns. It is a versatile tool used in various domains to improve decision-making and optimize processes. Understanding the data, selecting appropriate models, and evaluating model performance are critical components of effective predictive modeling.

Understanding the data, selecting appropriate models, and evaluating model performance are critical steps in sports analytics predictive modeling. In this example, we'll walk through these steps using a hypothetical dataset for predicting the outcome of soccer (football) matches.

```r
# Load required libraries
library(tidyverse)
library(caret)
library(randomForest)

# Load soccer match data (example dataset)
soccer_data <- read.csv("soccer_match_data.csv")

# Data Understanding
# Explore the dataset to understand its structure and characteristics
str(soccer_data)
summary(soccer_data)

# Data Preprocessing
# Assuming you have data cleaning, feature engineering, and target variable creation steps

# Selecting Appropriate Models
# Split the data into training and testing sets
set.seed(123)
train_indices <- createDataPartition(soccer_data$outcome, p = 0.8, list = FALSE)
train_data <- soccer_data[train_indices, ]
test_data <- soccer_data[-train_indices, ]

# Fit a Random Forest model as an example
rf_model <- randomForest(outcome ~ team1_strength + team2_strength + home_field_advantage,
            data = train_data, ntree = 500)

# Model Evaluation
# Predict the outcomes on the test set
rf_predictions <- predict(rf_model, newdata = test_data)

# Evaluate model performance using accuracy as an example metric
accuracy <- sum(rf_predictions == test_data$outcome) / length(test_data$outcome)

# Display the model performance
```

```
cat("Random Forest Model Accuracy:", accuracy, "\n")

# Additional evaluation metrics (e.g., confusion matrix, ROC curve) can be used

# Feature Importance
# Analyze feature importance to understand which factors influence match outcomes
importance <- importance(rf_model)
varImpPlot(importance, main = "Random Forest - Feature Importance")

# Model Interpretation
# Interpret the model's predictions and provide insights for decision-makers

# Model Deployment (if applicable)
# Deploy the model for real-time predictions or further analysis

# Continuous Model Monitoring
# Implement continuous monitoring to ensure the model's performance remains acceptable

# Conclusion and Recommendations
# Summarize findings and provide recommendations for improving team performance or strategies
```

In this example:

> We load hypothetical soccer match data and explore its structure and characteristics to gain an understanding of the dataset.

> Data preprocessing steps, including data cleaning, feature engineering, and target variable creation, are assumed.

> We select an appropriate model (Random Forest in this case) for predicting match outcomes.

> Model evaluation is performed using accuracy as an example metric, but other evaluation metrics like confusion matrices or ROC curves can be used. Feature importance analysis helps us understand which factors influence match outcomes the most.

> Model interpretation is crucial to provide insights for decision-makers in the sports industry.

If applicable, the model can be deployed for real-time predictions or further analysis.

Continuous model monitoring ensures that the model's performance remains acceptable over time.

The conclusion and recommendations section summarizes the findings and provides actionable recommendations based on the analysis.

These steps demonstrate the process of understanding data, selecting models, and evaluating their performance, which are vital in sports analytics predictive modeling.

Building predictive models for sports outcomes

Building predictive models for sports outcomes is an exciting application of predictive analytics that allows teams, analysts, and sports enthusiasts to make informed predictions about the results of sporting events. These models rely on historical data, performance metrics, and various factors specific to the sport to forecast game outcomes. Here's a step-by-step guide to building predictive models for sports outcomes:

1. Data Collection:
    - Gather historical data relevant to the sport and the specific outcomes you want to predict. This includes game results, player statistics, team rankings, venue information, and any other relevant factors. Data sources may include official sports databases, websites, and APIs.

2. Data Preprocessing:
    - Clean and preprocess the collected data to handle missing values, outliers, and inconsistencies. Ensure that the data is in a format suitable for modeling, with features and target variables clearly defined.

3. Feature Engineering:
    - Create meaningful features that capture the important aspects of the sport and teams. This may involve aggregating player statistics, calculating team averages, or deriving new variables that could influence game outcomes.

4. Model Selection:
    - Choose an appropriate predictive modeling technique based on the nature of the data and the problem. Common models for sports outcome prediction include logistic regression, decision trees, random forests, support vector machines, and neural networks.

5. Training and Testing Data:
   - Split the dataset into a training set and a testing (or validation) set. The training set is used to train the model, while the testing set is used to evaluate its performance.

6. Model Training:
   - Train the selected model using the training dataset. The model learns the patterns and relationships within the data to make predictions.

7. Performance Evaluation:
   - Assess the model's performance on the testing dataset using appropriate evaluation metrics. Common metrics for sports outcome prediction include accuracy, precision, recall, F1-score, and the area under the receiver operating characteristic curve (AUC-ROC).

8. Hyperparameter Tuning:
   - Optimize the model's hyperparameters to improve its predictive accuracy. This can involve techniques like grid search or random search.

9. Cross-Validation:
   - Implement cross-validation techniques (e.g., k-fold cross-validation) to ensure that the model's performance is robust and not overfitting to the training data.

10. Interpretability:
- Depending on the chosen model, consider methods for interpreting the model's predictions. Some models, like decision trees, offer straightforward interpretability.

11. Continuous Updating:
- Regularly update the model with new data to account for changes in team performance, player injuries, or other factors that may affect outcomes.

12. Model Deployment:
- Once satisfied with the model's performance, deploy it for making predictions about upcoming sporting events. This could be in the form of a web application, API, or reports for decision-makers.

13. Ethical Considerations:
- Ensure that the predictive model adheres to ethical standards, especially in sports betting or gambling contexts. Avoid biased or unfair predictions that may contribute to unethical practices.

14. Monitoring and Feedback:
- Continuously monitor the model's predictions and compare them to the actual outcomes. Collect feedback from users and stakeholders to make necessary improvements.

15. Interpretation and Action:
- Use the model's predictions to inform decisions related to sports betting, fantasy sports, team strategies, or player evaluations.

16. Transparency and Accountability:
- Be transparent about the model's limitations and assumptions. Ensure accountability in its use and address any concerns related to its fairness and accuracy.

Building predictive models for sports outcomes is a dynamic process that combines data science techniques with domain-specific knowledge of the sport. It allows sports enthusiasts to gain insights and make informed predictions while enhancing the overall fan experience.

Model evaluation and validation techniques

Model evaluation and validation techniques are critical steps in the development of predictive models to ensure their accuracy, generalizability, and reliability. These techniques help assess how well a model performs on unseen data and whether it can make accurate predictions. Here are common model evaluation and validation techniques:

1. Train-Test Split:
- Technique: Split the dataset into two subsets: a training set and a testing set. The model is trained on the training data and evaluated on the testing data.
- Purpose: To assess how well the model generalizes to new, unseen data.

2. k-Fold Cross-Validation:
- Technique: Divide the dataset into k subsets or "folds." Train and evaluate the model k times, using each fold as the testing set once while the remaining folds serve as the training data.
- Purpose: To assess model performance more robustly and reduce the impact of randomness in data splits.

3. Leave-One-Out Cross-Validation (LOOCV):
   - Technique: Similar to k-fold cross-validation, but with k equal to the number of data points in the dataset. In each iteration, one data point is used as the testing set, while the remaining points are used for training.
   - Purpose: To assess model performance when dealing with small datasets or when each data point is valuable.

4. Stratified Sampling:
   - Technique: When dealing with imbalanced datasets (where one class is much more frequent than another), ensure that the train-test split or cross-validation maintains the class distribution proportions in both subsets.
   - Purpose: To prevent models from being biased toward the majority class.

5. Evaluation Metrics:
   - Technique: Use appropriate evaluation metrics based on the nature of the problem. Common metrics include:
     - Classification: Accuracy, precision, recall, F1-score, ROC-AUC.
     - Regression: Mean Absolute Error (MAE), Mean Squared Error (MSE), Root Mean Squared Error (RMSE), R-squared (R2).
   - Purpose: To quantitatively measure the model's performance in making predictions.

6. Confusion Matrix:
   - Technique: In classification problems, create a confusion matrix to visualize true positives, true negatives, false positives, and false negatives.
   - Purpose: Provides a more detailed understanding of classification model performance, especially in imbalanced datasets.

7. Receiver Operating Characteristic (ROC) Curve:
   - Technique: Plot the ROC curve to visualize the trade-off between sensitivity (true positive rate) and specificity (true negative rate) by varying the classification threshold.
   - Purpose: Helps assess and compare the performance of binary classification models.

8. Precision-Recall Curve:
   - Technique: Plot the precision-recall curve to evaluate the precision-recall trade-off across different classification thresholds.
   - Purpose: Particularly useful when dealing with imbalanced datasets where precision and recall are more relevant than accuracy.

9. Model Comparison:
   - Technique: Compare the performance of different models using the same evaluation metrics to select the best-performing model.

- Purpose: To choose the model that best fits the problem and dataset.

10. Feature Importance Analysis:
- **Technique:** Assess the importance of individual features in influencing model predictions. Methods include feature importance scores or permutation importance.
- **Purpose:** Provides insights into which features contribute most to the model's predictions.

11. Bias and Fairness Assessment:
- **Technique:** Examine the model for potential bias or fairness issues, especially when making decisions that affect individuals or groups.
- **Purpose:** Ensure that models are not unfairly biased against certain demographics or protected classes.

12. Model Validation on Unseen Data:
- **Technique:** After selecting the final model, validate it on a completely new and unseen dataset to assess its real-world performance.
- **Purpose:** Ensures that the model performs well in practical applications and does not overfit to a specific dataset.

13. Error Analysis:
- **Technique:** Analyze the model's errors to understand common failure patterns and identify areas for model improvement.
- **Purpose:** Helps refine the model and data preprocessing steps.

Effective model evaluation and validation are essential to building reliable and accurate predictive models. These techniques help ensure that the model's performance is trustworthy and can make accurate predictions in real-world scenarios.

Below, I'll provide a comprehensive example in R, showcasing various techniques for model evaluation and validation using a hypothetical soccer (football) match prediction model.

```r
# Load required libraries
library(tidyverse)
library(caret)
library(randomForest)

# Load soccer match data (example dataset)
soccer_data <- read.csv("soccer_match_data.csv")

# Data Preprocessing
```

```r
# Assuming you have data cleaning, feature engineering, and target variable creation steps

# Split the data into training and testing sets
set.seed(123)
train_indices <- createDataPartition(soccer_data$outcome, p = 0.8, list = FALSE)
train_data <- soccer_data[train_indices, ]
test_data <- soccer_data[-train_indices, ]

# Model Training
# Fit a Random Forest model as an example
rf_model <- randomForest(outcome ~ team1_strength + team2_strength + home_field_advantage,
            data = train_data, ntree = 500)

# Model Evaluation

# 1. Accuracy
rf_predictions <- predict(rf_model, newdata = test_data)
accuracy <- sum(rf_predictions == test_data$outcome) / length(test_data$outcome)
cat("Accuracy:", accuracy, "\n")

# 2. Confusion Matrix
conf_matrix <- confusionMatrix(rf_predictions, test_data$outcome)
print(conf_matrix)

# 3. ROC Curve and AUC
library(ROCR)
roc_obj <- prediction(as.numeric(rf_predictions == "Win"), as.numeric(test_data$outcome == "Win"))
roc_perf <- performance(roc_obj, "tpr", "fpr")
auc <- performance(roc_obj, "auc")
plot(roc_perf, main = "ROC Curve")
cat("AUC:", unlist(slot(auc, "y.values")), "\n")

# 4. Precision, Recall, F1-Score
precision <- conf_matrix$byClass["Precision"]
recall <- conf_matrix$byClass["Recall"]
f1_score <- conf_matrix$byClass["F1"]
cat("Precision:", precision, "\n")
cat("Recall:", recall, "\n")
cat("F1-Score:", f1_score, "\n")
```

5. Cross-Validation
```
ctrl <- trainControl(method = "cv", number = 5)
cv_results <- train(outcome ~ team1_strength + team2_strength + home_field_advantage,
            data = train_data,
            method = "rf",
            trControl = ctrl)
print(cv_results)
```

Model Validation

1. Test on New Data (if available)
```
new_data <- read.csv("new_soccer_match_data.csv")
new_predictions <- predict(rf_model, newdata = new_data)
# Make predictions on new, unseen data
```

2. Model Monitoring
Implement continuous monitoring to ensure model performance over time.

Conclusion and Recommendations
Summarize the model evaluation results and provide actionable recommendations based on the analysis.

In this example:

We load hypothetical soccer match data and perform data preprocessing steps.

We split the data into training and testing sets.

We train a Random Forest model for match prediction using the training data. Model evaluation includes accuracy, confusion matrix, ROC curve, AUC, precision, recall, and F1-score.

We use cross-validation to assess model performance and generalization. Model validation involves testing on new, unseen data (if available) and implementing continuous model monitoring.

The conclusion and recommendations section summarizes the evaluation results and provides actionable insights.

These techniques demonstrate effective model evaluation and validation in sports analytics, helping ensure the reliability and accuracy of predictive models.

Predicting game outcomes and player performance

Predicting game outcomes and player performance in sports is a challenging yet fascinating application of predictive modeling and sports analytics. Whether for sports betting, fantasy sports, or team strategies, accurate predictions can provide valuable insights. Here's how you can approach predicting game outcomes and player performance:

Predicting Game Outcomes:

Data Collection: Gather historical game data, including team statistics, player statistics, venue information, and game results. Include data for both competing teams.

Feature Engineering: Create relevant features based on historical data, such as team win-loss records, player statistics (e.g., points scored, assists), head-to-head performance, home-field advantage, recent form, and more.

Target Variable: Define the target variable, which is typically binary (e.g., win/loss) or ordinal (e.g., win by margin). This represents the outcome you want to predict.

Model Selection: Choose an appropriate model for game outcome prediction. Common choices include logistic regression, decision trees, random forests, and support vector machines. Ensemble models like gradient boosting or neural networks can also be effective.

Training and Testing Data: Split the historical data into training and testing sets, ensuring that each set includes a balanced representation of different seasons and teams.

Model Training: Train the chosen model using the training dataset. The model learns the patterns and relationships in the data that lead to different game outcomes.

Model Evaluation: Evaluate the model's performance using appropriate classification metrics like accuracy, precision, recall, and F1-score. Additionally, consider using confusion matrices and ROC curves for a deeper understanding.

Hyperparameter Tuning: Optimize the model's hyperparameters to improve prediction accuracy. This can be done through techniques like grid search or random search.

Cross-Validation: Implement k-fold cross-validation to ensure that the model's performance is consistent across different subsets of the data.

Deployment: Once satisfied with the model's performance, deploy it to make predictions about upcoming games. Ensure that you regularly update the model with new data.

Ethical Considerations: Be mindful of the ethical implications of sports betting predictions, especially when sharing predictions with the public. Promote responsible gambling practices.

Predicting Player Performance:

Data Collection: Gather historical player data, including individual game statistics, player profiles, injury history, and performance metrics like Player Efficiency Rating (PER) or Win Shares.

Feature Engineering: Create relevant features, such as player averages, recent performance, opponent strength, playing position, and home-court advantage. Consider using advanced metrics like True Shooting Percentage (TS%), Effective Field Goal Percentage (eFG%), and Usage Rate (USG%).

Target Variable: Define the target variable for player performance, which could be points scored, rebounds, assists, or other individual statistics.

Model Selection: Choose an appropriate model for player performance prediction. Regression models (linear regression, ridge regression) and machine learning models (random forests, gradient boosting) are commonly used.

Training and Testing Data: Split the historical data into training and testing sets for model development and evaluation.

Model Training: Train the selected model using the training dataset. The model learns to predict player performance based on the provided features.

Model Evaluation: Evaluate the model's performance using appropriate regression metrics like Mean Absolute Error (MAE), Mean Squared Error (MSE), and Root Mean Squared Error (RMSE). R-squared (R2) can measure the model's goodness of fit.

Hyperparameter Tuning: Optimize the model's hyperparameters to improve prediction accuracy.

Cross-Validation: Implement k-fold cross-validation to assess how well the model generalizes to different datasets and seasons.

Player-Specific Models: Consider building individual models for different players or player positions, as player performance can vary significantly.

Continuous Updating: Regularly update player performance models with new data to account for player development and changing conditions.

Predicting game outcomes and player performance is a dynamic field in sports analytics that requires a deep understanding of the sport, relevant data, and predictive modeling techniques. Accurate predictions can have a significant impact on sports-related decisions and strategies.

In this example, we'll showcase how to predict NFL game outcomes and player performance using R. We'll use simplified data for demonstration purposes.

```
# Load required libraries
library(tidyverse)
library(caret)
library(randomForest)

# Load NFL game and player data (example datasets)
game_data <- read.csv("nfl_game_data.csv")
player_data <- read.csv("nfl_player_data.csv")

# Data Preprocessing
# Assuming you have data cleaning, feature engineering, and target variable creation steps

# Predicting NFL Game Outcomes

# Split the game data into training and testing sets
set.seed(123)
train_indices <- createDataPartition(game_data$outcome, p = 0.8, list = FALSE)
train_game_data <- game_data[train_indices, ]
test_game_data <- game_data[-train_indices, ]

# Fit a Random Forest model to predict game outcomes
rf_game_model <- randomForest(outcome ~ team1_strength + team2_strength + home_field_advantage,
            data = train_game_data, ntree = 500)

# Model Evaluation for Game Outcomes
```

```r
# Predict game outcomes on the test set
rf_game_predictions <- predict(rf_game_model, newdata = test_game_data)

# Evaluate model performance (e.g., accuracy, confusion matrix)
game_accuracy <- sum(rf_game_predictions == test_game_data$outcome) / length(test_game_data$outcome)
cat("Game Outcome Prediction Accuracy:", game_accuracy, "\n")

# Predicting NFL Player Performance

# Merge game data with player data to link player stats to specific games
merged_data <- merge(game_data, player_data, by.x = "player_id", by.y = "player_id")

# Split the merged data into training and testing sets
train_player_data <- merged_data[train_indices, ]
test_player_data <- merged_data[-train_indices, ]

# Fit a Random Forest model to predict player performance
rf_player_model <- randomForest(player_performance_metric ~ player_stat1 + player_stat2,
                data = train_player_data, ntree = 500)

# Model Evaluation for Player Performance

# Predict player performance on the test set
rf_player_predictions <- predict(rf_player_model, newdata = test_player_data)

# Evaluate model performance (e.g., RMSE, MAE, R-squared)
player_rmse <- sqrt(mean((rf_player_predictions - test_player_data$player_performance_metric)^2))
cat("Player Performance RMSE:", player_rmse, "\n")

# Conclusion and Recommendations

# Summarize the results, including game outcome predictions and player performance predictions.
# Provide recommendations for sports-related decisions and strategies based on the analysis.
```

In this example:

- We load hypothetical NFL game and player data and perform data preprocessing steps.

We split the game data into training and testing sets for predicting game outcomes and link player stats to specific games.

We train separate Random Forest models for predicting game outcomes and player performance.

Model evaluation includes metrics like accuracy for game outcomes and RMSE for player performance.

The conclusion and recommendations section summarizes the results and provides insights for sports-related decisions and strategies.

Keep in mind that this is a simplified example, and real-world NFL predictions involve more complex data, advanced modeling techniques, and domain-specific knowledge. Accurate predictions can significantly impact sports-related decisions and strategies, making sports analytics a dynamic and vital field.

Chapter 8: Player Draft and Recruitment

Draft strategy in different sports

Draft strategies in different sports vary significantly due to the unique characteristics, rules, and objectives of each sport's draft system. Here's a general overview of draft strategies in some popular sports:

1. NFL (National Football League):
    - Objective: In the NFL Draft, teams aim to acquire talented players who can strengthen their rosters, particularly in positions of need.
    - Strategy:
        - Priority Positioning: Teams often prioritize positions critical to their success, such as quarterback, offensive line, or pass rushers.
        - Trade Opportunities: Teams may trade up or down in the draft order to secure specific players or acquire additional picks.
        - Value-Based Drafting: Teams evaluate players based on their perceived value, taking into account their skills, potential, and how they fit into the team's system.
        - Balancing Immediate and Future Needs: Teams must balance drafting for immediate impact with building for the future.
        - Player Character and Health: Consideration of a player's character, injury history, and work ethic is crucial in draft decisions.

2. NBA (National Basketball Association):
    - Objective: NBA teams aim to draft players who can contribute to their success on the court, either immediately or in the near future.
    - Strategy:
        - Talent vs. Need: Teams may prioritize the best available player (BPA) over addressing positional needs, especially in the early rounds.
        - International Prospects: The NBA drafts international players who may not have extensive exposure to American basketball. Scouting and evaluation of international talent are essential.
        - Trades: Draft night trades involving picks and players are common. Teams may trade up to select a specific player or acquire assets for the future.

- Potential vs. Proven Performance: Teams must assess the potential of young, unproven players versus the track record of more experienced college players.

3. MLB (Major League Baseball):
   - Objective: MLB teams aim to replenish their farm systems with talented young players who can eventually contribute at the major league level.
   - Strategy:
     - Scouting and Player Development: MLB teams invest heavily in scouting and player development to identify and nurture young talent.
     - Risk vs. Reward: Teams evaluate players based on their potential upside, considering factors like hitting power, pitching velocity, and defensive skills.
     - Positional Depth: Some teams prioritize drafting players to address positional weaknesses in their minor league systems.
     - High School vs. College: Teams must decide between drafting high school prospects with high ceilings or college players with more experience.
     - Analytics: Advanced analytics play a growing role in assessing player performance and projecting future success.

4. NHL (National Hockey League):
   - Objective: NHL teams seek to draft skilled players who can contribute to their team's success, with a focus on positions like forwards, defensemen, and goaltenders.
   - Strategy:
     - Skill Development: Hockey teams emphasize player development, as young prospects often need time to mature physically and refine their skills.
     - Importance of Goaltending: Goaltenders are a critical part of NHL success, and teams may draft and develop goaltending prospects with care.
     - Positional Depth: Teams consider the depth of talent available in specific positions when making draft decisions.
     - International Prospects: The NHL drafts players from around the world, so scouting and assessing international talent are essential.
     - Draft Age: Hockey players are often drafted at a younger age, and teams must project how players will develop over time.

5. MLS (Major League Soccer):
   - Objective: MLS teams aim to acquire young soccer talent to improve their squads and develop players who can contribute on the field.
   - Strategy:

- International Signings: MLS teams often sign international players, but the SuperDraft provides an opportunity to select top collegiate talent.
- Homegrown Players: Some MLS teams prioritize developing players from their youth academies and may focus less on the SuperDraft.
- Positional Needs: Teams consider their positional needs when drafting players, especially in defense, midfield, and forward positions.
- Trading Picks: Teams may trade draft picks to acquire established players or assets.

In all sports, successful draft strategies require a combination of scouting, data analysis, and strategic decision-making. Teams must balance the need for immediate impact with long-term planning to build competitive rosters. Additionally, draft strategies can evolve over time based on changes in team dynamics and league rules.

Using data to evaluate draft prospects

Evaluating draft prospects using data is a fundamental aspect of modern sports analytics. Data-driven scouting and analysis can provide teams with valuable insights into the potential performance and impact of prospective players. Here's a step-by-step guide on how data is used to evaluate draft prospects:

1. Data Collection:
   - Gather comprehensive data on draft prospects, including their game statistics, physical attributes, injury history, and any relevant off-field information. This data can be obtained from college or amateur leagues, combine events, and scouting reports.

2. Data Preprocessing:
   - Clean and preprocess the collected data to handle missing values, outliers, and inconsistencies. Ensure that the data is in a structured format suitable for analysis.

3. Feature Engineering:
   - Create meaningful features or variables that can help assess a prospect's performance and potential. Examples include player efficiency ratings (PER), win shares, player comparisons, and position-specific metrics.

4. Performance Metrics:
   - Develop performance metrics specific to the sport and position. For example, in basketball, you might consider metrics like shooting efficiency, rebounding rates, assists per game, and defensive metrics like steal and block rates.

5. Comparative Analysis:
   - Compare a prospect's performance metrics with those of other players in the same draft class or historical drafts. This helps assess where a prospect stands relative to peers.

6. Statistical Analysis:
   - Use statistical techniques to identify trends, patterns, and outliers in the data. Statistical tests can help determine whether a prospect's performance is statistically significant and predictive.

7. Predictive Modeling:
   - Build predictive models that estimate a prospect's future performance or impact in the professional league. Common modeling techniques include regression analysis, decision trees, random forests, and machine learning algorithms.

8. Positional Analysis:
   - Tailor the evaluation to the specific requirements of the player's position. For example, in basketball, point guards and centers have different skill sets and responsibilities, so their evaluations may focus on different aspects of their game.

9. Combine and Workout Data:
   - Include data from scouting combines, workouts, and interviews. Combine results like 40-yard dash times, vertical jumps, and bench press numbers can provide additional insights into a prospect's physical attributes and potential.

10. Injury and Health Analysis:
- Consider a prospect's injury history and medical assessments. This information is crucial for assessing a player's durability and long-term potential.

11. Player Comparisons:
- Compare prospects to current or former professional players who share similar attributes or playing styles. This can help teams envision a prospect's role in their system.

12. Ethical Considerations:
- Be mindful of the ethical considerations involved in collecting and using data on prospects. Respect privacy and ensure that data is used responsibly and transparently.

13. Scouting Reports:
- Combine quantitative data-driven analysis with qualitative scouting reports and expert opinions. Scouts' observations and insights are valuable complements to statistical analysis.

14. Final Evaluation:
- Synthesize all available data, analysis, and evaluations to create a comprehensive assessment of a prospect's potential. This assessment guides draft decisions.

15. Draft Strategy:
- Incorporate data-driven evaluations into the team's draft strategy, which may involve prioritizing prospects with the highest predicted impact or addressing specific positional needs.

16. Continuous Improvement:
- Continuously update and refine the evaluation process based on new data, research, and emerging analytics techniques.

Data-driven evaluation of draft prospects has become a crucial component of professional sports team operations. It helps teams make more informed decisions, reduces the risk of drafting underperforming players, and maximizes the potential for building competitive rosters.

In this example, we'll use R to demonstrate how to evaluate NFL draft prospects based on historical player data and predict their potential performance.
For this demonstration, we'll assume you have access to a dataset containing historical NFL player statistics and draft information.

```r
# Load required libraries
library(tidyverse)
library(caret)
library(randomForest)

# Load historical NFL player data (example dataset)
player_data <- read.csv("nfl_player_data.csv")

# Data Preprocessing
# Assuming you have data cleaning, feature engineering, and target variable creation steps

# Selecting Draft Prospects

# Filter the dataset to include only draft-eligible players
draft_eligible_players <- player_data %>%
```

```r
  filter(draft_eligible == 1)

# Split the data into training and testing sets
set.seed(123)
train_indices <- createDataPartition(draft_eligible_players$draft_position, p = 0.8, list = FALSE)
train_data <- draft_eligible_players[train_indices, ]
test_data <- draft_eligible_players[-train_indices, ]

# Feature Selection
# Choose relevant player attributes as features for the model
selected_features <- c("college_stats", "combine_performance", "position")

# Fit a Random Forest model to predict draft position
rf_draft_model <- randomForest(draft_position ~ ., data = train_data[, selected_features], ntree = 500)

# Model Evaluation

# Predict draft positions on the test set
rf_draft_predictions <- predict(rf_draft_model, newdata = test_data[, selected_features])

# Evaluate model performance (e.g., RMSE, MAE, R-squared)
draft_rmse <- sqrt(mean((rf_draft_predictions - test_data$draft_position)^2))
cat("Draft Position Prediction RMSE:", draft_rmse, "\n")

# Conclusion and Recommendations

# Summarize the results, including draft position predictions for prospects.
# Provide recommendations for teams based on the analysis, considering draft strategy and player potential.
```

In this example:

We load hypothetical NFL player data, which includes information about college stats, combine performance, and draft eligibility.

We filter the dataset to include only draft-eligible players.
The data is split into training and testing sets, with the goal of predicting the draft position.

We select relevant player attributes as features for the model.

We train a Random Forest model to predict the draft position based on the selected features.

Model evaluation is performed using RMSE, which measures the accuracy of draft position predictions.

The conclusion and recommendations section summarizes the results and provides insights for teams on draft strategy and player potential.

Case studies of successful draft selections

Successful draft selections are often celebrated in sports as they can have a profound impact on a team's performance and success. Here are a few case studies of notable successful draft selections in different sports:

1. NBA - Michael Jordan (1984 NBA Draft):
   - Drafted by: Chicago Bulls (3rd overall)
   - Impact: Michael Jordan is widely regarded as one of the greatest basketball players of all time. He led the Chicago Bulls to six NBA championships and won five NBA Most Valuable Player (MVP) awards during his career.

2. NFL - Tom Brady (2000 NFL Draft):
   - Drafted by: New England Patriots (199th overall)
   - Impact: Tom Brady, drafted in the sixth round, became a seven-time Super Bowl champion and is considered one of the greatest quarterbacks in NFL history. He has numerous MVP awards and records to his name.

3. MLB - Mike Trout (2009 MLB Draft):
   - Drafted by: Los Angeles Angels (25th overall)
   - Impact: Mike Trout quickly emerged as one of the best players in baseball. He has won multiple American League MVP awards and is known for his combination of power, speed, and defensive prowess.

4. NHL - Sidney Crosby (2005 NHL Draft):
   - Drafted by: Pittsburgh Penguins (1st overall)
   - Impact: Sidney Crosby, the first overall pick, has lived up to expectations, winning multiple Stanley Cups with the Penguins and numerous MVP awards. He is considered one of the best hockey players of his generation.

5. MLS - Carlos Vela (2005 MLS SuperDraft):
   - Drafted by: Chivas USA (1st overall)

- Impact: Carlos Vela, as the top pick in the 2005 MLS SuperDraft, has had a successful career both in MLS and internationally. He became one of the league's top goal-scorers during his time with LAFC.

6. NBA - Giannis Antetokounmpo (2013 NBA Draft):
    - Drafted by: Milwaukee Bucks (15th overall)
    - Impact: Giannis Antetokounmpo, drafted in the middle of the first round, developed into an NBA superstar. He has won multiple MVP awards and led the Bucks to an NBA championship.

7. NFL - Patrick Mahomes (2017 NFL Draft):
    - Drafted by: Kansas City Chiefs (10th overall)
    - Impact: Patrick Mahomes, drafted in the first round, quickly rose to prominence as one of the most exciting and successful quarterbacks in the NFL. He won the Super Bowl and Super Bowl MVP in his second season as a starter.

8. MLB - Albert Pujols (1999 MLB Draft):
    - Drafted by: St. Louis Cardinals (13th round)
    - Impact: Albert Pujols, a 13th-round pick, had a remarkable career, becoming one of the greatest hitters in baseball history. He won multiple MVP awards and helped lead the Cardinals to two World Series championships.

These case studies highlight that successful draft selections can come from various rounds and positions in the draft. Talent evaluation, scouting, and player development play crucial roles in identifying prospects with the potential for long and impactful careers in professional sports.

Free agency and recruitment analytics

Free agency and recruitment analytics play a significant role in professional sports, helping teams make informed decisions about signing or acquiring players. These analytics involve the use of data-driven techniques to assess a player's value, fit within a team's system, and potential impact on performance. Here's how free agency and recruitment analytics are utilized:

1. Player Valuation:
    - Determine a player's value based on their past performance, statistics, and contributions to their previous teams. Metrics like Player Efficiency Rating (PER) in basketball or Wins Above Replacement (WAR) in baseball are commonly used.

2. Market Analysis:
   - Assess the market demand for specific positions or skill sets. Teams analyze their own needs and identify positions that require reinforcement through free agency or recruitment.

3. Salary Cap Management:
   - Evaluate how a player's potential contract aligns with the team's salary cap constraints. Analyze the financial impact of a signing on the team's ability to remain competitive in the long term.

4. Fit and Style of Play:
   - Examine how a player's skills and style of play align with the team's existing roster and playing strategy. Ensure that a new signing complements the team's overall game plan.

5. Injury History:
   - Analyze a player's injury history and assess the risk associated with potential health issues. This is crucial for making long-term investments in players.

6. Age and Career Trajectory:
   - Consider a player's age and projected career trajectory. Younger players may have more potential for improvement, while older players may bring experience and leadership.

7. Advanced Metrics:
   - Use advanced analytics to evaluate a player's impact beyond traditional statistics. For example, in basketball, analytics may assess a player's impact on spacing, ball movement, or defensive efficiency.

8. Risk Assessment:
   - Assess the risk associated with a player's character, off-field behavior, or potential disruptions to team chemistry. This can be especially relevant in team sports where cohesion is essential.

9. Performance Projections:
   - Utilize predictive modeling to project a player's future performance with the team. This can help estimate the player's potential contribution to wins and losses.

10. Negotiation Strategy:
- Develop negotiation strategies based on the analytics to ensure teams get the best possible deal for the player they want to acquire.

11. Scouting Reports:
- Combine quantitative analytics with qualitative scouting reports and expert opinions to get a comprehensive view of a player's potential impact.

12. Long-Term Planning:
- Consider the long-term impact of player signings on the team's competitive window. Analyze whether a signing aligns with the team's championship aspirations.

13. Post-Signing Evaluation:
- Continuously monitor and evaluate the performance of players acquired through free agency or recruitment. Assess whether the signing met expectations and adjust team strategies accordingly.

Analytics in free agency and recruitment help teams optimize their roster, allocate resources efficiently, and increase their chances of success in professional sports. By leveraging data and analytics, teams can make more informed and strategic decisions in player acquisition.

In this example, we'll use R to demonstrate how to analyze and make decisions regarding free agency and recruitment in the context of professional basketball (NBA).

For this demonstration, we'll assume you have access to a dataset containing historical NBA player statistics, salary information, and team data.

```r
# Load required libraries
library(tidyverse)
library(caret)
library(randomForest)

# Load NBA player data (example dataset)
player_data <- read.csv("nba_player_data.csv")

# Data Preprocessing
# Assuming you have data cleaning, feature engineering, and target variable creation steps

# Analyzing Free Agency and Recruitment

# Select relevant features for recruitment decisions
selected_features <- c("player_stats", "age", "position", "salary")

# Split the data into training and testing sets
set.seed(123)
```

```r
train_indices <- createDataPartition(player_data$recruited, p = 0.8, list = FALSE)
train_data <- player_data[train_indices, ]
test_data <- player_data[-train_indices, ]

# Fit a Random Forest model to predict recruitment decisions
rf_recruitment_model <- randomForest(recruited ~ ., data = train_data[, selected_features], ntree = 500)

# Model Evaluation

# Predict recruitment decisions on the test set
rf_recruitment_predictions <- predict(rf_recruitment_model, newdata = test_data[, selected_features])

# Evaluate model performance (e.g., accuracy, confusion matrix)
recruitment_accuracy <- sum(rf_recruitment_predictions == test_data$recruited) / length(test_data$recruited)
cat("Recruitment Decision Accuracy:", recruitment_accuracy, "\n")

# Analyzing Salary Allocation

# Select relevant features for salary allocation decisions
selected_salary_features <- c("player_stats", "age", "position", "recruited")

# Fit a Random Forest model to predict player salaries
rf_salary_model <- randomForest(salary ~ ., data = train_data[, selected_salary_features], ntree = 500)

# Model Evaluation for Salary Allocation

# Predict player salaries on the test set
rf_salary_predictions <- predict(rf_salary_model, newdata = test_data[, selected_salary_features])

# Evaluate model performance (e.g., RMSE, MAE, R-squared)
salary_rmse <- sqrt(mean((rf_salary_predictions - test_data$salary)^2))
cat("Salary Prediction RMSE:", salary_rmse, "\n")

# Conclusion and Recommendations

# Summarize the results, including recruitment decisions and salary allocation predictions.
```

Provide recommendations for teams on player acquisition and salary allocation based on the analysis.

In this example:

> We load hypothetical NBA player data, which includes information about player statistics, age, position, salary, and recruitment decisions.
>
> We select relevant features for recruitment and salary allocation decisions.
>
> The data is split into training and testing sets, with the goal of predicting recruitment decisions and player salaries.
>
> We train Random Forest models for both recruitment decisions and salary allocation based on the selected features.
>
> Model evaluation includes accuracy for recruitment decisions and RMSE for salary predictions.
>
> The conclusion and recommendations section summarizes the results and provides insights for teams on player acquisition and salary allocation.

Chapter 9: Fan Engagement and Marketing

Analyzing fan behavior and preferences

Analyzing fan behavior and preferences is a critical aspect of sports marketing, fan engagement, and business strategy for sports organizations. Understanding what fans want and how they interact with a team or sport can help organizations tailor their offerings, improve the fan experience, and grow their fan base. Here's how fan behavior and preferences are analyzed:

1. Surveys and Questionnaires:
   - Conduct surveys and questionnaires among fans to gather information about their preferences, interests, and demographics. Questions can cover topics such as favorite players, game attendance frequency, merchandise preferences, and reasons for supporting the team.

2. Social Media Monitoring:
   - Monitor social media platforms to track fan sentiment, engagement, and conversations related to the team or sport. Analyze comments, likes, shares, and hashtags to understand what fans are talking about and how they feel.

3. Ticketing and Attendance Data:
   - Analyze ticket sales data and attendance records to identify trends in game attendance. This can help determine which games are most popular, which sections of the stadium are preferred, and factors affecting attendance fluctuations.

4. Website and App Analytics:
   - Use website and mobile app analytics to track user behavior. Understand which content is most popular, which pages fans visit the most, and how long they spend on the site or app.

5. Fan Engagement Metrics:
   - Measure fan engagement metrics, such as the number of social media followers, email subscribers, and interactions with fan content (e.g., likes, comments, shares). Analyze trends to see what content resonates most with fans.

6. Email Surveys and Feedback:
   - Send email surveys to fans to collect feedback after specific events or promotions. This provides insights into the fan experience and areas for improvement.

7. Focus Groups:
   - Conduct focus group sessions with a cross-section of fans to gain qualitative insights into their preferences, attitudes, and motivations. These sessions can provide deeper context for quantitative data.

8. Customer Relationship Management (CRM) Systems:
   - Use CRM systems to track fan interactions and preferences. This helps personalize communication and marketing efforts based on individual fan profiles.

9. Merchandise Sales Data:
   - Analyze merchandise sales data to identify top-selling items and design new products that align with fan preferences. Seasonal trends and promotions can also be derived from this data.

10. Game-Day Experience Feedback:
- Gather feedback from fans about their in-stadium or arena experience. This can include surveys, mobile apps for feedback, and post-game interviews.

11. Geographical Analysis:
- Analyze geographical data to understand where fans are located and how to target specific regions or demographics. This can inform marketing and outreach efforts.

12. Season Ticket Holder Insights:
- Focus on season ticket holders as they are often the most committed fans. Analyze their behavior, preferences, and feedback to enhance their experience and retention.

13. Trend Analysis:
- Keep an eye on industry trends, such as changes in viewing habits, emerging technologies, and fan engagement tactics employed by other sports organizations.

14. A/B Testing:
- Experiment with different marketing campaigns, content strategies, and fan engagement initiatives. A/B testing can help identify which approaches resonate best with fans.

15. Data Integration:
- Integrate data from various sources, such as social media, ticketing, merchandise, and website analytics, to create a comprehensive view of fan behavior and preferences.

Analyzing fan behavior and preferences is an ongoing process, as fan preferences can evolve over time. Sports organizations must use data-driven insights to continuously adapt and enhance their strategies to create a more engaging and enjoyable experience for their fans.

Data-driven marketing strategies

Data-driven marketing strategies involve using data and analytics to make informed decisions about how to reach and engage with target audiences effectively. These strategies enable businesses and organizations to optimize their marketing efforts, personalize messaging, and improve overall campaign performance. Here are key components and steps in implementing data-driven marketing strategies:

1. Data Collection:
   - Gather data from various sources, including customer interactions, website traffic, social media, email campaigns, and offline channels. Ensure data quality and compliance with data privacy regulations (e.g., GDPR, CCPA).

2. Customer Segmentation:
   - Use data to segment your audience into distinct groups based on demographics, behavior, preferences, or purchase history. Segmentation allows for more personalized and targeted marketing efforts.

3. Customer Profiling:
   - Create detailed customer profiles or personas based on data, including age, gender, location, interests, and buying habits. Understand what motivates your customers and how they interact with your brand.

4. Predictive Analytics:
   - Utilize predictive analytics to forecast future customer behavior and preferences. This can help identify potential leads, churn risks, and upsell opportunities.

5. Content Personalization:
   - Customize marketing content, including emails, website content, and product recommendations, based on customer data and preferences. Personalization increases engagement and conversion rates.

6. Marketing Automation:
   - Implement marketing automation tools to streamline repetitive tasks, such as email marketing, lead nurturing, and social media posting. Automation helps deliver timely and relevant messages.

7. A/B Testing:
   - Conduct A/B tests to compare different marketing approaches, such as email subject lines, ad creatives, or landing page designs. Analyze the results to optimize campaigns.

8. Attribution Modeling:
   - Use attribution models to understand the impact of various marketing channels on conversions. Determine which channels and touchpoints contribute most to your goals.

9. Real-Time Analytics:
   - Monitor and analyze marketing campaigns in real-time to make adjustments as needed. Identify trends, opportunities, and issues promptly.

10. Customer Journey Mapping:
- Create customer journey maps to visualize the stages and touchpoints a customer goes through when interacting with your brand. Tailor marketing messages to align with these stages.

11. Customer Lifetime Value (CLV):
- Calculate CLV to assess the long-term value of a customer. Use this information to allocate marketing resources more effectively and prioritize high-value customer segments.

12. Data Visualization:
- Use data visualization tools to present data in a clear and visually appealing manner. Dashboards and reports help marketers and decision-makers quickly grasp insights.

13. Customer Feedback Analysis:
- Analyze customer feedback from surveys, reviews, and social media. Identify trends and sentiments to improve products, services, and marketing strategies.

14. Data Security and Privacy:
- Prioritize data security and comply with data privacy regulations to protect customer information and maintain trust.

15. Continuous Optimization:
- Continuously assess and optimize marketing strategies based on data-driven insights. Be agile in making changes to campaigns and tactics.

16. ROI Measurement:
- Calculate the return on investment (ROI) for marketing campaigns. Analyze which campaigns generate the highest returns and allocate resources accordingly.

17. Customer Retention Strategies:
- Use data to identify at-risk customers and implement retention strategies. Personalized offers, loyalty programs, and proactive customer support can help retain valuable customers.

Data-driven marketing strategies enable businesses to adapt to changing consumer behavior, improve customer experiences, and maximize the effectiveness of marketing investments. By leveraging data and analytics, organizations can stay competitive and meet the evolving demands of their target audience.

In this example, we'll use R to demonstrate how to analyze fan behavior and preferences to create data-driven marketing strategies for a professional sports team, such as an NBA team.

For this demonstration, we'll assume you have access to a dataset containing historical fan engagement data, ticket sales, and marketing campaign results.

```r
# Load required libraries
library(tidyverse)
library(ggplot2)

# Load fan engagement data (example dataset)
fan_data <- read.csv("nba_fan_data.csv")

# Data Preprocessing
# Assuming you have data cleaning, feature engineering, and target variable creation steps

# Analyzing Fan Behavior and Preferences

# Explore fan data to understand fan behavior and preferences
summary(fan_data)

# Calculate fan engagement metrics (e.g., attendance, social media interactions)
fan_data <- fan_data %>%
```

```
    mutate(total_interactions = social_media_likes + social_media_comments + social_media_shares)

# Visualize fan engagement metrics
ggplot(fan_data, aes(x = game_attendance, y = total_interactions)) +
  geom_point() +
  labs(title = "Fan Engagement vs. Game Attendance",
      x = "Game Attendance",
      y = "Total Social Media Interactions")

# Segment fans based on behavior and preferences (e.g., high-engagement fans, ticket buyers)
high_engagement_fans <- fan_data %>%
  filter(total_interactions >= quantile(total_interactions, 0.75))
ticket_buyers <- fan_data %>%
  filter(ticket_purchased == 1)

# Analyze the characteristics of high-engagement fans and ticket buyers
high_engagement_summary <- summary(high_engagement_fans)
ticket_buyer_summary <- summary(ticket_buyers)

# Data-Driven Marketing Strategies

# Develop personalized marketing campaigns targeting high-engagement fans and ticket buyers
# Tailor marketing messages, offers, and promotions based on their preferences and behavior

# Evaluate the effectiveness of marketing campaigns by tracking ticket sales and fan engagement
# Analyze ROI (Return on Investment) for each campaign

# Continuously monitor fan behavior and preferences to adapt marketing strategies
# Use A/B testing and customer segmentation for ongoing optimization

# Conclusion and Recommendations

# Summarize the results, including insights from fan behavior and preferences analysis.
# Provide recommendations for data-driven marketing strategies to improve fan engagement and ticket sales.
```

In this example:

We load hypothetical fan engagement data, which includes information about game attendance, social media interactions, and ticket purchases.

We preprocess the data, including data cleaning and feature engineering.

We explore fan behavior and preferences using summary statistics and visualizations.

Fan engagement metrics are calculated, and the relationship between game attendance and social media interactions is visualized.

We segment fans based on behavior and preferences, such as high-engagement fans and ticket buyers.

Data-driven marketing strategies are developed, focusing on personalized marketing campaigns tailored to fan segments.

The effectiveness of marketing campaigns is evaluated by tracking ticket sales and fan engagement, including ROI analysis.

Continuous monitoring and optimization of marketing strategies are emphasized, including A/B testing and customer segmentation.

The conclusion and recommendations section summarizes the results and provides insights for data-driven marketing strategies to enhance fan engagement and ticket sales.

Enhancing the fan experience through analytics

Enhancing the fan experience through analytics is a critical focus for sports organizations and entertainment venues. By leveraging data and analytics, these organizations can better understand their fans, tailor offerings, and create memorable experiences that keep fans engaged and loyal. Here's how analytics can be used to enhance the fan experience:

1. Personalized Marketing and Engagement:
    - Fan Segmentation: Use data to segment fans based on demographics, preferences, behavior, and attendance history. Tailor marketing messages, promotions, and content to each segment.

- Recommendation Engines: Implement recommendation algorithms to suggest relevant products, tickets, and experiences to fans based on their past interactions and preferences.
- Behavioral Tracking: Track fan interactions with websites, apps, and social media. Analyze this data to deliver personalized content and offers in real-time.

2. Ticketing and Seating Optimization:
   - Dynamic Pricing: Implement dynamic pricing strategies that adjust ticket prices based on factors like demand, opponent, and seat location to offer fans better value.
   - Seat Selection: Use analytics to help fans choose the best seats based on their preferences, such as proximity to the action or specific amenities.

3. In-Stadium Experience:
   - Mobile Apps: Develop mobile apps that provide fans with real-time information, wayfinding, and interactive features like ordering concessions or merchandise for delivery to their seats.
   - Wait-Time Prediction: Use data to predict wait times at restrooms, concession stands, and other amenities, allowing fans to plan their visits more efficiently.
   - Augmented Reality (AR) and Virtual Reality (VR): Integrate AR and VR experiences to provide fans with immersive experiences, such as virtual tours of the stadium or interactive games during breaks.

4. Enhanced Services:
   - Customer Service Chatbots: Implement AI-powered chatbots to assist fans with inquiries, ticket purchases, and service requests in real-time.
   - Fan Feedback Analysis: Analyze fan feedback from surveys, social media, and other channels to identify areas for improvement in stadium facilities, services, and fan engagement initiatives.

5. Game Enhancements:
   - Live Stats and Analytics: Provide fans with real-time statistics, player tracking data, and insights on the game through digital displays and mobile apps.
   - Fan Engagement Zones: Create interactive fan zones within the stadium where fans can participate in games, contests, and social media activities related to the event.

6. Loyalty Programs:
   - Fan Loyalty Points: Implement loyalty programs that reward fans for attending games, making purchases, and engaging with the team's brand. Offer exclusive benefits and experiences to loyal fans.

7. Post-Event Engagement:
   - Surveys and Feedback: Collect post-event feedback to understand fan satisfaction and areas for improvement. Use this data to refine future fan experiences.
   - Content Delivery: Continue engaging with fans through post-game content, highlights, behind-the-scenes footage, and exclusive interviews to keep the excitement alive.

8. Data Security and Privacy:
   - Prioritize the security and privacy of fan data to build trust and ensure compliance with data protection regulations.

Enhancing the fan experience through analytics is an ongoing process. Organizations must continuously collect and analyze data, stay updated on technological advancements, and respond to changing fan preferences and expectations. By delivering personalized and engaging experiences, sports organizations can foster stronger connections with fans and build a loyal and enthusiastic fan base.

Case studies of successful fan engagement campaigns

Successful fan engagement campaigns are crucial for sports organizations and teams to build a strong and loyal fan base, increase attendance, and boost revenue. Here are some case studies of successful fan engagement campaigns:

1. Manchester City FC - #Cityzens:
   - Objective: Manchester City FC aimed to strengthen its global fan community and increase fan engagement.
   - Campaign: The club launched the #Cityzens program, which provided fans with exclusive access to content, matchday experiences, and rewards. Fans could earn points by attending games, interacting on social media, and participating in surveys and quizzes.
   - Results: The #Cityzens program saw significant growth in fan engagement and loyalty. Members enjoyed exclusive benefits, including meeting players and attending behind-the-scenes events.

2. Seattle Seahawks - 12 Flag Raising:
   - Objective: The Seattle Seahawks aimed to enhance the gameday experience and engage their passionate fan base, known as the "12s."
   - Campaign: Before each home game, the Seahawks raise the "12 Flag" at their stadium. This tradition involves selecting a local celebrity or fan to raise the flag, generating excitement and anticipation among fans.

- Results: The 12 Flag raising ceremony has become an iconic part of the Seahawks' gameday experience, fostering a strong sense of community and pride among fans.

3. Denver Broncos - "Be a Bronco for a Day" Sweepstakes:
    - Objective: The Denver Broncos sought to engage fans and create excitement around the team.
    - Campaign: The team launched the "Be a Bronco for a Day" sweepstakes, allowing fans to enter for a chance to spend a day at the Broncos' facility, meet players, and attend practice.
    - Results: The campaign generated significant fan participation and excitement. Winners enjoyed an unforgettable experience and shared their stories on social media, further boosting fan engagement.

4. Chicago Bulls - #BullsNation:
    - Objective: The Chicago Bulls aimed to connect with fans and build a strong online community.
    - Campaign: The team launched the #BullsNation campaign, encouraging fans to use the hashtag and share their Bulls-related experiences, photos, and stories on social media.
    - Results: The campaign successfully united fans under the #BullsNation banner, creating a sense of belonging and camaraderie among supporters. It also generated user-generated content and expanded the team's digital presence.

5. Major League Baseball - "Opening Day" Celebrations:
    - Objective: Major League Baseball (MLB) sought to celebrate the start of the baseball season and engage fans.
    - Campaign: MLB hosts "Opening Day" celebrations with special events, parades, and fan engagement activities across the league. Fans are encouraged to attend games or participate in local festivities.
    - Results: Opening Day generates excitement and enthusiasm among fans, marking the official start of the baseball season. It serves as a unifying moment for baseball enthusiasts across the country.

6. Philadelphia Eagles - Eagles Autism Challenge:
    - Objective: The Philadelphia Eagles aimed to engage fans while supporting a charitable cause.
    - Campaign: The Eagles launched the Eagles Autism Challenge, a fundraising event where fans could participate in various activities, including a bike ride, run/walk, or virtual challenge. Proceeds went to autism research and care.
    - Results: The campaign not only engaged fans but also raised substantial funds for autism research and support. Fans felt a sense of pride in contributing to a meaningful cause supported by their favorite team.

These case studies highlight the power of fan engagement campaigns to strengthen the bond between sports organizations and their fans. Successful campaigns often involve a combination of exclusivity, community-building, social media, and meaningful experiences that resonate with fans' passions and values.

Chapter 10: Ethics and Fair Play in Sports Analytics

Ethical considerations in sports analytics

Ethical considerations in sports analytics are crucial to ensure the responsible and fair use of data and analytics techniques. Sports organizations, analysts, and researchers must uphold ethical standards to protect the privacy and rights of athletes, maintain the integrity of sports competitions, and promote transparency. Here are key ethical considerations in sports analytics:

1. Data Privacy and Consent:
- Athlete Consent: Athletes should be informed about the collection, use, and sharing of their personal and performance data. Obtaining informed consent from athletes is essential, especially when collecting sensitive information.
- Data Security: Safeguard athlete data through secure storage and transmission to prevent unauthorized access, breaches, or data leaks.

2. Fair Play and Competition:
- Competitive Integrity: Ensure that analytics and data-driven insights do not compromise the integrity of sports competitions. Avoid tactics that may provide unfair advantages or undermine fair play.
- Performance Enhancing Drugs: Ethical concerns arise when using analytics to detect doping or performance-enhancing substances. Uphold the principles of fairness, confidentiality, and due process in anti-doping efforts.

3. Transparency and Accountability:
- Transparency: Be transparent about the use of analytics in sports. Provide clear explanations of how data is collected, analyzed, and used to make decisions.
- Accountability: Establish clear lines of accountability for decisions made based on analytics. Responsible parties should be identifiable and responsible for the outcomes.

4. Data Bias and Discrimination:
   - Data Bias: Be vigilant in identifying and mitigating biases in data and algorithms. Biased data can perpetuate discrimination or unfair practices, such as player recruitment or performance evaluation.
   - Equal Opportunity: Ensure that analytics do not discriminate against athletes based on race, gender, age, or other protected characteristics. Use data responsibly to promote equal opportunities.

5. Fan Privacy:
   - Fan Data Protection: Sports organizations must also consider the privacy of their fans. Data collected from ticket sales, merchandise purchases, or engagement on digital platforms should be handled responsibly and in compliance with privacy regulations.

6. Ethical Use of Injury Data:
   - Injury Privacy: Respect the privacy of injured athletes. Medical and injury data should only be shared with relevant medical staff and used for the athlete's well-being.
   - Avoid Exploitation: Do not exploit injury data for sensationalism or media coverage that could harm the athlete's reputation or well-being.

7. Youth Sports and Data Usage:
   - Protecting Minors: When collecting data on youth athletes, ensure that the rights and privacy of minors are safeguarded. Parental consent may be required for data collection and analysis.

8. Ethical Reporting:
   - Accurate Reporting: Report analytics findings accurately and without exaggeration or distortion. Misleading reporting can misinform fans, teams, and decision-makers.
   - Ethical Reporting of Injuries: When reporting on athlete injuries, respect the athlete's privacy and adhere to guidelines provided by sports organizations.

9. Ethical Responsibility of Analysts:
   - Analyst Training: Sports analysts should be trained in ethical standards and responsible data handling. Ethical conduct is a shared responsibility among all professionals in the field.

10. Informed Decision-Making:
- Athlete Welfare: Prioritize athlete welfare and well-being in decision-making processes based on analytics. Ethical considerations should extend beyond performance to include health and safety.

Sports organizations and analysts must continuously assess their practices and policies to align with evolving ethical standards and regulations. Ethical sports analytics not only preserves the integrity of sports but also ensures that athletes, fans, and stakeholders are treated with fairness and respect.

Fair play and data integrity

Fair play and data integrity are essential principles in sports analytics and data-driven decision-making. Ensuring fair play and data integrity helps maintain the credibility of sports competitions and the ethical use of data in sports. Here are key considerations related to fair play and data integrity in sports:

1. Data Accuracy:
   - Ensure that the data collected for analysis, whether related to athlete performance or game statistics, is accurate and reliable. Inaccurate data can lead to unfair assessments and decisions.

2. Consistency and Standardization:
   - Use consistent data collection methods and standards across different teams and competitions to maintain fairness. Standardized metrics and measurements enable fair comparisons.

3. Data Transparency:
   - Be transparent about the sources of data and the methods used for data collection, processing, and analysis. Transparency builds trust among athletes, teams, and fans.

4. Data Security:
   - Protect data from unauthorized access, tampering, or manipulation. Data security measures are essential to maintain data integrity.

5. Fair Data Sharing:
   - When sharing data between teams or organizations, ensure that both parties have equal access to the same data sets. Unfair advantage can result from unequal access to data.

6. Anti-Doping and Fair Play:
   - Use data analytics to detect and prevent doping or unfair practices that undermine fair play in sports. Uphold the principles of fair competition and ethical standards in anti-doping efforts.

7. Privacy and Consent:
- Respect the privacy and data rights of athletes. Obtain informed consent for data collection, and ensure that athlete data is used only for legitimate purposes.

8. Data Bias Mitigation:
- Be vigilant in identifying and mitigating biases in data and algorithms. Data-driven decisions should not perpetuate discrimination or unfairly disadvantage certain athletes or teams.

9. Ethical Reporting:
- Report analytics findings ethically and accurately. Misleading or biased reporting can lead to unfair judgments or misperceptions.

10. Governance and Oversight:
- Establish governance structures and oversight mechanisms to ensure that data collection and analysis adhere to fair play and ethical principles. This may include ethics committees or compliance officers.

11. Continuous Monitoring:
- Continuously monitor data practices and data sources to detect and address any potential breaches of data integrity or fair play.

12. Data Education:
- Educate athletes, coaches, and other stakeholders about the role of data analytics in sports and the ethical standards that must be maintained.

Fair play and data integrity go hand in hand to create a level playing field in sports. Sports organizations, analysts, and regulators must work together to uphold these principles and promote ethical behavior in the collection, analysis, and use of data in sports. By doing so, they can maintain the integrity of sports competitions and the trust of athletes and fans alike.

Privacy concerns and data security

Privacy concerns and data security are paramount when it comes to sports analytics and the handling of athlete and fan data. Protecting the privacy of individuals and ensuring the security of data is not only a legal obligation but also essential for maintaining trust and ethical standards. Here are key considerations related to privacy and data security in sports analytics:

1. Athlete Data Privacy:
   - Athlete Consent: Obtain informed consent from athletes before collecting and using their personal and performance data. Clearly explain the purposes for which the data will be used.
   - Data Minimization: Collect only the data that is necessary for the intended purpose. Avoid excessive data collection.
   - Privacy Policies: Develop and communicate privacy policies that outline how athlete data will be handled, who will have access to it, and how long it will be retained.

2. Fan Data Privacy:
   - Fan Consent: When collecting fan data for marketing or engagement purposes, seek consent and provide options for fans to control the data they share.
   - Opt-In/Opt-Out: Allow fans to opt in or opt out of data collection and marketing communications. Respect their preferences.

3. Data Security:
   - Encryption: Encrypt sensitive data during transmission and storage to protect it from unauthorized access or breaches.
   - Access Control: Implement strict access controls to ensure that only authorized personnel can access and modify data. Use role-based access control (RBAC) to manage permissions.
   - Regular Security Audits: Conduct regular security audits and vulnerability assessments to identify and address potential weaknesses in your data infrastructure.

4. Compliance with Regulations:
   - GDPR and CCPA: Ensure compliance with data protection regulations such as the General Data Protection Regulation (GDPR) and the California Consumer Privacy Act (CCPA) when handling data of individuals residing in regions covered by these regulations.
   - HIPAA: If handling medical or health-related data, adhere to the Health Insurance Portability and Accountability Act (HIPAA) regulations.

5. Data Retention and Deletion:
   - Define clear data retention policies and delete data when it is no longer needed for its intended purpose. This includes athlete and fan data.

6. Third-Party Services:
   - Vet third-party vendors and services for their data security practices. Ensure that they comply with your privacy and security standards.

7. Incident Response Plan:
   - Develop an incident response plan to address data breaches or security incidents promptly. Communicate breaches to affected individuals and authorities as required by law.

8. Training and Awareness:
   - Train employees, athletes, and fans about data privacy and security best practices. Create awareness about the importance of protecting personal information.

9. Transparency:
   - Be transparent with athletes and fans about how their data is used, who has access to it, and how it is secured. Transparency builds trust.

10. Data Ethics:
- Promote ethical data practices, including responsible data collection, analysis, and use. Avoid practices that could lead to unfair advantages or discriminatory outcomes.

Privacy concerns and data security are ongoing responsibilities in sports analytics. Organizations should regularly review and update their data privacy and security practices to align with evolving regulations and emerging threats. By prioritizing privacy and security, sports organizations can maintain the trust of athletes, fans, and stakeholders while leveraging data for insights and improvements.

Responsible use of analytics in sports

The responsible use of analytics in sports involves ethical considerations and best practices to ensure that data-driven decisions and technologies are used in a fair, transparent, and ethical manner. Here are guidelines for the responsible use of analytics in sports:

1. Privacy and Data Protection:
   - Obtain Informed Consent: Seek informed consent from athletes and fans before collecting and using their data. Clearly communicate the purposes for which data will be used.
   - Data Minimization: Collect only the data that is necessary for the intended purpose. Avoid excessive data collection.
   - Data Security: Implement robust data security measures, including encryption and access controls, to protect data from unauthorized access or breaches.
   - Compliance: Ensure compliance with data protection regulations such as GDPR, CCPA, and HIPAA when handling athlete and fan data.

2. Fair Play and Competition:
   - Maintain Competitive Integrity: Ensure that analytics and data-driven insights do not compromise the fairness and integrity of sports competitions. Avoid tactics that may provide unfair advantages.
   - Anti-Doping: Use analytics to detect and prevent doping or performance-enhancing substances, adhering to the principles of fairness and due process.

3. Transparency:
   - Data Transparency: Be transparent about the sources of data, data collection methods, and the algorithms used for analysis. Transparency builds trust among athletes, teams, and fans.
   - Explainability: Ensure that analytics models and decisions are explainable. Athletes and stakeholders should understand how decisions are made based on data.

4. Ethical Reporting:
   - Accurate Reporting: Report analytics findings accurately and without exaggeration or distortion. Misleading reporting can misinform fans, teams, and decision-makers.

5. Data Bias Mitigation:
   - Identify and Mitigate Bias: Be vigilant in identifying and mitigating biases in data and algorithms. Data-driven decisions should not perpetuate discrimination or unfairly disadvantage certain athletes or teams.

6. Data Ethics:
   - Promote Responsible Practices: Encourage responsible data practices, including ethical data collection, analysis, and use. Avoid practices that could lead to unfair advantages or discriminatory outcomes.
   - Ethical Use of Injury Data: Respect the privacy of injured athletes. Medical and injury data should only be shared with relevant medical staff and used for the athlete's well-being.

7. Governance and Oversight:
   - Establish governance structures and oversight mechanisms to ensure that data collection and analysis adhere to ethical and responsible standards. This may include ethics committees or compliance officers.

8. Athlete and Fan Engagement:
   - Involve athletes and fans in the data-driven process. Allow them to have a voice in how data is used and provide feedback mechanisms.

9. Continuous Monitoring:
- Continuously assess data practices and data sources to detect and address potential breaches of ethics and responsible use.

10. Education and Awareness:
- Educate athletes, coaches, analysts, and other stakeholders about the role of data analytics in sports and the ethical standards that must be maintained.

Responsible use of analytics in sports requires a commitment to ethical standards, transparency, and fairness. By adhering to these principles, sports organizations can harness the power of data analytics to enhance performance, fan engagement, and the overall sports experience while upholding the integrity of the game.

Chapter 11: Case Studies in Sports Analytics

Real-world case studies from various sports

Here are some real-world case studies from various sports that showcase the practical applications and impact of analytics:

1. Moneyball (Baseball):
   - Case: The Oakland Athletics, as depicted in Michael Lewis's book "Moneyball," used advanced analytics to identify undervalued players and build a competitive team on a limited budget. They focused on player statistics like on-base percentage and slugging percentage to make data-driven decisions.
   - Impact: The Athletics' data-driven approach revolutionized player scouting and recruitment in baseball. It demonstrated how analytics could be used to find hidden talent and compete against teams with larger budgets.

2. Leicester City's Premier League Win (Soccer/Football):
   - Case: In the 2015-2016 English Premier League season, Leicester City, a relatively small club, won the league against all odds. They used analytics to assess player performance, develop tactical strategies, and manage player fatigue.
   - Impact: Leicester City's triumph demonstrated that data analytics could level the playing field in soccer. Their win is often cited as one of the most remarkable achievements in the history of the sport.

3. Golden State Warriors (Basketball):
   - Case: The Golden State Warriors, an NBA team, have used analytics extensively to optimize player rotations, shot selection, and defensive strategies. They employed a combination of on-court player tracking data and video analysis.
   - Impact: The Warriors became one of the most successful teams in the NBA, winning multiple championships. Their analytics-driven approach influenced other teams to adopt similar strategies.

4. Mercedes-AMG Petronas Formula One Team (Formula 1):
   - Case: The Mercedes-AMG Petronas Formula One Team uses data analytics to fine-tune their car's performance. They collect and analyze data from various sensors on the car to make real-time adjustments during races.
   - Impact: Mercedes has dominated Formula 1 in recent years, in part due to their data-driven approach. They showcase how analytics can give teams a competitive edge in a high-speed, precision sport.

5. FC Barcelona's Sports Science Institute (Soccer/Football):
   - Case: FC Barcelona's Sports Science Institute uses analytics to monitor players' physical and physiological data, such as heart rate, speed, and distance covered during matches and training sessions. This data helps in injury prevention and performance optimization.
   - Impact: FC Barcelona's focus on sports science and analytics has contributed to their sustained success in both domestic and international competitions.

6. New England Patriots (American Football):
   - Case: The New England Patriots, an NFL team, have a reputation for using analytics to make strategic decisions. They analyze opponent tendencies, player performance, and situational factors to optimize their game plans.
   - Impact: The Patriots have won multiple Super Bowl championships, and their analytics-driven approach is recognized as a key factor in their success.

7. Seattle Sounders (Soccer/Football):
   - Case: The Seattle Sounders of Major League Soccer (MLS) use analytics to optimize ticket pricing. They adjust ticket prices based on factors like opponent strength, day of the week, and historical attendance data.
   - Impact: The Sounders' data-driven ticket pricing strategy has led to increased attendance and revenue while ensuring fans have access to more affordable tickets.

These case studies demonstrate the diverse applications of analytics in various sports and how data-driven approaches can lead to competitive advantages, improved performance, and enhanced fan engagement.

In-depth analysis of notable games or seasons

In-depth analysis of notable games or seasons in sports provides valuable insights into the strategies, performances, and historical significance of these events. Let's delve into a few notable games and seasons across different sports:

1. Super Bowl LI (American Football):
   - Game Overview: Super Bowl LI, played in 2017, featured the New England Patriots vs. the Atlanta Falcons. The game is notable for the Patriots' historic comeback from a 28-3 deficit to win in overtime.
   - Analysis: The game showcased the importance of resilience, adaptability, and effective game planning in football. The Patriots adjusted their strategy, and quarterback Tom Brady's performance was instrumental in their victory.

2. 2016 NBA Finals (Basketball):
   - Series Overview: The 2016 NBA Finals featured the Cleveland Cavaliers vs. the Golden State Warriors. The Cavaliers overcame a 3-1 series deficit to win their first NBA championship.
   - Analysis: This series emphasized the impact of individual brilliance, teamwork, and LeBron James' exceptional performance. It also highlighted the significance of mental toughness in high-pressure situations.

3. 2019 Cricket World Cup Final (Cricket):
   - Game Overview: The 2019 Cricket World Cup final was contested between England and New Zealand. The match ended in a dramatic tie, leading to a Super Over, which also ended in a tie. England won based on the number of boundaries scored.
   - Analysis: This game exemplified the unpredictable nature of cricket and the importance of rules and regulations. The boundary count tiebreaker sparked debates about fairness and alternative tiebreak methods.

4. 2018 FIFA World Cup (Soccer/Football):
   - Season Overview: The 2018 FIFA World Cup held in Russia featured several notable moments, including the rise of underdog teams like Croatia and the success of France.
   - Analysis: This World Cup illustrated the global nature of soccer and how smaller nations can compete at the highest level. It also emphasized the role of team dynamics and strategy in achieving success.

5. 2020 Tokyo Olympics (Various Sports):
   - Season Overview: The 2020 Tokyo Olympics, postponed to 2021 due to the COVID-19 pandemic, featured a wide range of sports and memorable moments, including standout performances by athletes like Simone Biles and Eliud Kipchoge.
   - Analysis: The Olympics showcased the resilience of athletes and the global unity through sports, even during challenging times. It highlighted the importance of adaptability and preparation for unforeseen circumstances.

6. 2019 Wimbledon Men's Final (Tennis):
   - Game Overview: The 2019 Wimbledon Men's Final between Roger Federer and Novak Djokovic was one of the longest and most closely contested Wimbledon finals in history, lasting nearly five hours.
   - Analysis: This match highlighted the mental and physical endurance required in tennis and the ability to perform under extreme pressure. It showcased the rivalry between two tennis legends.

Analyzing notable games and seasons provides a deeper understanding of the dynamics of sports, the strategies employed by athletes and teams, and the impact of individual and collective performances. These events often become historic moments in the world of sports and leave a lasting legacy.

Lessons learned from successful analytics projects

Successful analytics projects in sports offer valuable lessons that can be applied to similar endeavors in other domains. Here are some key lessons learned from successful sports analytics projects:

1. Clear Objectives and KPIs:
   - Define clear project objectives and key performance indicators (KPIs) from the outset. A well-defined goal helps guide the project and measure its success.

2. Data Quality Matters:
   - Ensure the quality, accuracy, and completeness of the data used for analysis. Data integrity is foundational to reliable insights.

3. Cross-Disciplinary Teams:
   - Build cross-disciplinary teams that include data analysts, domain experts, and decision-makers. Collaboration between experts in different fields leads to more informed decisions.

4. Understand the Domain:
   - Gain a deep understanding of the specific domain or sport you're analyzing. Contextual knowledge is essential for meaningful insights.

5. Iterative Approach:
   - Embrace an iterative approach to analytics. Be prepared to adjust your methods and models as you gather more data and insights.

6. Use Advanced Techniques:
- Don't shy away from advanced analytics techniques, such as machine learning and predictive modeling, when appropriate. These methods can reveal hidden patterns and trends.

7. Real-Time Analysis:
- When possible, implement real-time analysis to make timely decisions during games or events. Quick insights can provide a competitive advantage.

8. Ethical Considerations:
- Adhere to ethical standards and privacy regulations. Protect the privacy of individuals and ensure data is used responsibly.

9. Effective Communication:
- Communicate findings and insights effectively to decision-makers, coaches, or players. Clear communication is essential for actionable results.

10. Test Hypotheses:
- Formulate hypotheses and test them rigorously. Use data to validate or refute assumptions and hypotheses.

11. Evaluate Impact:
- Assess the impact of analytics on decision-making and performance. Ensure that the project delivers tangible benefits.

12. Scalability:
- Consider the scalability of your analytics solution. Can it be applied to different teams, seasons, or sports?

13. Continuous Learning:
- Encourage a culture of continuous learning and improvement. Stay updated on the latest analytics techniques and technologies.

14. Change Management:
- Be prepared for resistance to change. Implement strategies to manage and overcome resistance from stakeholders.

15. Flexibility:
- Be flexible in your approach and willing to adapt to changing circumstances. Sports environments are dynamic, and flexibility is key to success.

16. Celebrate Successes:
- Celebrate the successes and milestones achieved through analytics projects. Recognition can motivate teams and stakeholders.

17. Share Knowledge:
- Share knowledge and best practices within your organization or community. Collaboration benefits the entire sports analytics community.

18. Long-Term Perspective:
- Maintain a long-term perspective. Analytics is an ongoing process, and long-term commitment is often necessary to see significant results.

These lessons underscore the importance of a holistic approach to sports analytics, combining technical expertise, domain knowledge, ethical considerations, and effective communication. Successful sports analytics projects demonstrate the potential of data-driven decision-making to improve performance, optimize strategies, and gain a competitive edge in the world of sports.

Challenges and failures in sports analytics

Challenges and failures in sports analytics are essential aspects of the field's growth and development. Learning from these setbacks helps refine approaches and improve the quality of insights. Here are some common challenges and failures in sports analytics:

1. Data Quality and Availability:
   - Challenge: Incomplete or inaccurate data can hinder analysis and lead to incorrect conclusions.
   - Failure: Relying on unreliable or incomplete data can result in flawed strategies and decisions.

2. Resistance to Analytics:
   - Challenge: Coaches, players, or team management may resist adopting analytics-driven strategies due to traditional approaches or skepticism.
   - Failure: Failure to convince stakeholders of the value of analytics can lead to missed opportunities for improvement.

3. Ethical Concerns:
   - Challenge: Balancing the use of personal and sensitive player data with ethical considerations can be complex.
   - Failure: Mishandling player data can lead to privacy breaches, legal issues, and damaged reputations.

4. Overreliance on Data:
   - Challenge: Overreliance on analytics without considering other factors like intuition or experience can lead to suboptimal decisions.

- Failure: Focusing solely on data can ignore the human element of sports, including team dynamics and player psychology.

5. Misinterpretation of Data:
   - Challenge: Data can be misinterpreted or misunderstood, leading to incorrect conclusions.
   - Failure: Misinterpreting data can result in misguided strategies and poor decision-making.

6. Lack of Domain Expertise:
   - Challenge: Analysts without a deep understanding of the specific sport may struggle to make meaningful insights.
   - Failure: Inaccurate or irrelevant insights can arise from a lack of domain expertise.

7. Inadequate Technology and Resources:
   - Challenge: Limited resources or outdated technology can impede data collection, analysis, and reporting.
   - Failure: Inadequate resources can hinder the implementation of analytics strategies.

8. Difficulty in Predicting Outcomes:
   - Challenge: Sports are inherently unpredictable, and some events cannot be accurately predicted with analytics alone.
   - Failure: Failing to acknowledge the inherent unpredictability of sports can lead to unrealistic expectations.

9. Player and Team Variability:
   - Challenge: Player and team performance can vary significantly from one season or game to another.
   - Failure: Overestimating the stability of player and team performance can lead to inaccurate predictions and strategies.

10. Lack of Data Standardization:
- **Challenge:** Data from different sources may lack standardization, making it challenging to integrate and analyze.
- **Failure:** Inconsistent data can result in errors and hinder cross-team or cross-sport comparisons.

11. Model Overfitting:
- **Challenge:** Overfitting occurs when models are too complex and fit the training data perfectly but perform poorly on new data.
- **Failure:** Overfitted models may produce unreliable predictions and insights.

12. Fan and Media Pressure:
- **Challenge:** High expectations from fans and media can put pressure on teams to adopt analytics quickly, potentially leading to rushed decisions.
- **Failure:** Rushed implementation of analytics can result in suboptimal strategies.

13. Lack of Cultural Shift:
- **Challenge:** Achieving a cultural shift towards data-driven decision-making within sports organizations can be slow and challenging.
- **Failure:** A failure to foster a culture of analytics can limit the impact of data-driven strategies.

Despite these challenges and failures, sports analytics continues to evolve and make a significant impact on the sports industry. Learning from these experiences, adapting strategies, and continuously improving the use of analytics is essential for its long-term success.

Chapter 12: Future Trends in Sports Analytics

Emerging technologies in sports analytics

Emerging technologies are continuously reshaping the landscape of sports analytics, offering new opportunities for teams, athletes, and organizations to gain insights and competitive advantages. Here are some of the emerging technologies in sports analytics:

**1. Player Tracking Technology: Advances in player tracking technology have revolutionized sports analytics. This technology uses sensors, GPS, and cameras to capture real-time data on player movements, positions, and performance metrics. It enables detailed analysis of player behavior and team strategies.

**2. Wearable Technology: Athletes can wear smart devices like fitness trackers and smartwatches to monitor various physiological parameters, including heart rate, sleep patterns, and activity levels. These wearables provide valuable data for assessing player health, recovery, and performance.

**3. Biomechanics Analysis: Biomechanics technology uses motion capture and sensors to analyze athletes' movements and techniques. It helps identify areas for improvement in technique and reduces the risk of injury.

**4. Computer Vision and AI: Computer vision and artificial intelligence (AI) are used to analyze video footage of games and practices. These technologies can automatically track player movements, identify patterns, and provide insights into tactics and strategies.

**5. Machine Learning and Predictive Analytics: Machine learning algorithms are increasingly used for player performance prediction, injury risk assessment, and game outcome prediction. These models analyze historical data to make informed predictions.

**6. Virtual Reality (VR) and Augmented Reality (AR): VR and AR are used for immersive training experiences and game simulations. Athletes can practice in virtual environments, enhancing their decision-making and skills.

**7. Sports Biometrics: Biometric sensors measure physiological data such as heart rate, body temperature, and hydration levels. This information helps monitor athletes' health and optimize their performance.

**8. Advanced Data Visualization: Advanced data visualization tools and software enable teams and analysts to present complex data in easily understandable formats. Interactive dashboards and heatmaps are examples of these tools.

**9. Blockchain Technology: Blockchain is used for secure and transparent data storage, particularly for sports betting and fantasy sports platforms. It ensures the integrity of data and transactions.

**10. 5G Technology: The rollout of 5G networks improves data transfer speeds and enables real-time data analytics and streaming. This is particularly useful for live game analysis and fan engagement.

**11. Quantum Computing: Although still in its infancy, quantum computing has the potential to revolutionize sports analytics by solving complex optimization problems and simulations more efficiently.

**12. Bioinformatics: Bioinformatics tools analyze genetic and genomic data to understand how genetics affect an athlete's performance, injury risk, and recovery.

**13. Drones: Drones capture aerial footage of games and practices, offering unique perspectives for analysis and scouting. They are also used for monitoring outdoor sports venues.

**14. IoT Sensors: Internet of Things (IoT) sensors are embedded in equipment, clothing, and venues to collect data on player and ball movement, environmental conditions, and more.

**15. Natural Language Processing (NLP): NLP is used for sentiment analysis of fan feedback, social media commentary, and press coverage. It helps teams gauge public opinion and adapt strategies accordingly.

These emerging technologies are transforming the world of sports analytics, providing teams, athletes, and organizations with unprecedented insights, enhancing fan engagement, and optimizing performance on and off the field. As these technologies continue to evolve, sports analytics will remain at the forefront of innovation in sports.

The role of machine learning and AI

Machine learning (ML) and artificial intelligence (AI) play significant roles in sports analytics, offering powerful tools for data analysis, prediction, and decision-making. Here's an overview of the roles of ML and AI in sports analytics:

1. Player Performance Prediction:
   - ML models can predict player performance based on historical data and various performance metrics. This helps teams make informed decisions about starting lineups, substitutions, and player development.

2. Injury Risk Assessment:
   - ML and AI algorithms analyze player health data, training loads, and injury histories to assess the risk of injuries. Teams can use this information to develop injury prevention strategies and optimize player recovery.

3. Game Outcome Prediction:
   - ML models can predict game outcomes by analyzing team statistics, player performance, and situational factors. These predictions are valuable for sports betting, fantasy sports, and team strategy.

4. Tactical Analysis:
   - AI can analyze video footage of games to identify patterns, tactics, and strategies used by teams and players. Coaches use this information to adjust game plans and exploit opponents' weaknesses.

5. Player Tracking:
   - ML algorithms process data from player tracking technology, such as GPS and RFID sensors, to monitor player movements and positions in real-time. This data helps teams understand player behavior, positioning, and fitness levels.

6. Performance Optimization:
   - ML can optimize player training regimens by analyzing performance data and customizing workouts to individual player needs. This leads to improved player development and performance.

7. Fan Engagement:
   - AI-driven chatbots and virtual assistants enhance fan engagement by providing real-time updates, answering fan queries, and delivering personalized content.

8. Scouting and Recruitment:
   - ML and AI help teams identify promising talent through data-driven scouting. These technologies analyze player statistics, performance data, and potential to assist in recruitment decisions.

9. Game Strategy:
   - ML algorithms assess opponent strategies and make recommendations for in-game adjustments. Coaches use this information to adapt their tactics and exploit weaknesses.

10. Data Processing and Insights:
- ML automates data preprocessing, reducing the time required to clean and prepare data for analysis. It can also uncover hidden insights and correlations within large datasets.

11. Natural Language Processing (NLP):
- NLP techniques are used to analyze and interpret text-based data, such as fan feedback, social media commentary, and news articles. This helps teams gauge public sentiment and adapt their strategies accordingly.

12. Sports Betting and Fantasy Sports:
- AI algorithms provide real-time odds, player recommendations, and predictive insights for sports betting platforms and fantasy sports applications.

13. Real-time Analytics:
- AI systems process real-time data during games or events, allowing teams to make immediate decisions based on live data streams.

Machine learning and AI have become integral components of sports analytics, enabling teams, athletes, and organizations to gain a competitive edge, make data-driven decisions, and enhance fan experiences. As technology continues to advance, the role of ML and AI in sports analytics will only become more significant.

Predictions for the future of sports analytics

The future of sports analytics holds several exciting possibilities and trends that are likely to shape the field in the coming years. Here are some predictions for the future of sports analytics:

1. Increased Adoption of AI and Machine Learning:
   - AI and machine learning will become even more prevalent in sports analytics. Advanced algorithms will analyze complex data sets, providing deeper insights into player performance, injury prevention, and game strategies.

2. Enhanced Player Tracking Technology:
   - Player tracking technology will continue to evolve, offering more granular data on player movements, biometrics, and real-time analytics. Wearable devices and sensors will become standard equipment for athletes, providing continuous performance monitoring.

3. Improved Injury Prevention:
   - Sports organizations will increasingly focus on data-driven injury prevention strategies. AI will predict injury risks, and personalized training and recovery plans will be developed for each athlete to minimize the risk of injuries.

4. Fan Engagement Innovation:
   - Analytics-driven fan engagement will reach new heights. Virtual reality (VR), augmented reality (AR), and personalized content will offer immersive fan experiences. Fans will have access to real-time stats, augmented live broadcasts, and interactive game simulations.

5. Customized Training Programs:
   - Player development will become highly personalized. ML algorithms will analyze individual player data to create tailored training programs, optimizing performance and skill development.

6. Advanced Data Visualization:
   - Data visualization tools will continue to improve, making complex analytics more accessible to coaches, players, and fans. Interactive dashboards and augmented reality visualizations will enhance data interpretation.

7. eSports Analytics Growth:
   - eSports analytics will expand as competitive video gaming gains popularity. Data-driven insights will help eSports teams refine strategies and enhance player performance.

8. Expanded Use of Biometric Data:
   - Biometric sensors and data will be integrated more extensively into sports analytics. Metrics like heart rate variability, sleep patterns, and cognitive performance will be used to assess player readiness and mental acuity.

9. Quantum Computing Impact:
   - As quantum computing matures, it will be employed for advanced simulations, optimization problems, and predictive modeling. This may lead to breakthroughs in sports analytics by solving complex problems faster.

10. Ethical Considerations:
- Ethical considerations surrounding data privacy and transparency will become increasingly important. Sports organizations will need to prioritize responsible data usage and maintain fan trust.

11. Greater Cross-Sport Integration:
- Analytics techniques and best practices will be shared across different sports. Lessons learned in one sport may lead to innovations in others.

12. Data Ecosystems:
- Sports organizations will build comprehensive data ecosystems that include not only player and game data but also fan data, marketing analytics, and business intelligence. This holistic approach will drive better decision-making across all aspects of sports operations.

The future of sports analytics promises to be dynamic and transformative, with technology-driven insights influencing player performance, injury management, fan experiences, and even the strategies employed by coaches and teams. As analytics tools and methodologies continue to advance, sports organizations that embrace data-driven decision-making will gain a competitive edge in their respective sports.

Opportunities and challenges in the field

The field of sports analytics presents numerous opportunities and challenges for sports organizations, athletes, analysts, and fans. Understanding these can help stakeholders maximize the benefits and address potential obstacles:

Opportunities in Sports Analytics:

Performance Enhancement: Sports analytics can lead to improved player performance through data-driven training programs, injury prevention, and real-time feedback.

Competitive Advantage: Teams and organizations can gain a competitive edge by leveraging analytics to optimize strategies, player recruitment, and in-game decision-making.

Fan Engagement: Analytics-driven fan experiences, such as real-time stats, fantasy sports, and augmented reality applications, enhance fan engagement and loyalty.

Injury Prevention: Advanced analytics can help identify injury risks and improve player health management, extending careers and reducing medical costs.

Talent Identification: Data analysis aids in identifying and recruiting promising talent, both in traditional sports and emerging fields like eSports.

Data-Driven Sponsorships: Sponsors can use analytics to target specific fan demographics and measure the ROI of their investments in sports events and teams.

Revenue Generation: Sports analytics can lead to revenue optimization by improving ticket sales, merchandise marketing, and concession strategies.

Improved Training: Coaches and trainers can develop more effective training regimens by tailoring programs to individual athletes based on their data profiles.

Globalization of Sports: Analytics can help sports organizations expand their reach to global audiences, fostering international fan bases and market opportunities.

Challenges in Sports Analytics:

Data Quality and Availability: Access to high-quality, comprehensive data can be limited, leading to incomplete or inaccurate analyses.

Resistance to Change: Coaches, players, and organizations may resist adopting analytics, citing traditional approaches or skepticism about new methods.

Privacy and Ethical Concerns: Handling personal data, especially in medical and injury analysis, raises privacy and ethical considerations.

Interpretation Complexity: Misinterpreting data or relying solely on analytics can lead to poor decisions if the human element and context are overlooked.

Technological Limitations: Developing and maintaining analytics infrastructure, including hardware and software, can be resource-intensive and challenging.

Data Security: Protecting sensitive player and fan data from breaches and cyberattacks is a significant concern.

Fan Data Usage: Balancing fan engagement with data privacy is crucial, as overzealous data collection can lead to backlash.

Overemphasis on Metrics: Overreliance on statistics may undervalue intangible aspects of sports, such as team chemistry and leadership.

Regulatory Compliance: Staying compliant with data protection regulations, such as GDPR and CCPA, can be complex, especially for international sports organizations.

Predictive Accuracy: Predicting sports outcomes and player performance is inherently challenging due to the unpredictability of sports events.

Navigating these opportunities and challenges in sports analytics requires a balanced approach, ethical considerations, and a commitment to responsible data practices. As the field continues to evolve, stakeholders must adapt to technological advancements and changing landscapes to unlock its full potential.

Chapter 13: Getting Started with R in Sports Analytics

Setting up an R environment for sports analytics

Setting up an R environment for sports analytics involves installing the R programming language, RStudio (an integrated development environment), and relevant packages for data analysis and visualization. Here are the steps to set up an R environment for sports analytics:

1. Install R:
   - Go to the R project website and choose a CRAN mirror near your location.
   - Download the R installer for your operating system (Windows, macOS, or Linux).
   - Follow the installation instructions for your specific operating system.

2. Install RStudio:
   - RStudio is a popular integrated development environment (IDE) for R that provides a user-friendly interface for coding and data analysis.
   - Go to the RStudio download page and download the appropriate installer for your operating system.
   - Follow the installation instructions to install RStudio.

3. Open RStudio:
   - After installing RStudio, open it from your computer's applications or programs menu.

4. Install R Packages:
   - R packages are collections of functions and data sets that extend R's capabilities. To perform sports analytics, you'll need specific packages.
   - Open RStudio and use the R console or RScript to install packages. For example, you can install the "dplyr" package for data manipulation using the following command:

install.packages("dplyr")

- Install other relevant packages, such as "ggplot2" for data visualization and "SportsAnalytics" for sports-specific data (if available).

5. Load Packages:
   - Once installed, load the packages you want to use in your R script. Use the library() function to load a package. For example:

library(dplyr)
library(ggplot2)

6. Import Data:
   - To perform sports analytics, you'll need data. You can import data from various sources like CSV files, databases, or APIs using functions like read.csv(), read.table(), or specialized packages depending on the data source.

7. Analyze and Visualize Data:
   - Use R's data manipulation and visualization packages to analyze and visualize sports data. For example, you can use the "dplyr" package for data manipulation and "ggplot2" for creating interactive visualizations.

8. Learn R:
   - To effectively use R for sports analytics, it's important to learn the language and its packages. There are many online resources, courses, and tutorials available for learning R, including those focused on sports analytics.

9. Explore Sports-Specific Packages:
   - Depending on your specific area of sports analytics (e.g., basketball, soccer, baseball), explore packages tailored to that sport. For instance, there are packages like "nba_api" for NBA data or "SportsAnalytics" for general sports data.

10. Practice and Experiment:
- The more you work on sports analytics projects in R, the more proficient you'll become. Practice and experiment with different datasets and analysis techniques to gain experience.

Remember that sports analytics can encompass a wide range of topics, including player performance analysis, team strategy, injury prediction, and fan engagement. Building a strong foundation in R and regularly updating your knowledge of sports-specific analytics will help you excel in this field.

Basic R programming for data analysis

To get started with basic R programming for data analysis, you'll need to understand some fundamental concepts and commands. Here, I'll provide an overview of the key elements you'll commonly use in R for data analysis:

1. Data Types:
    - Numeric: Represents real numbers, such as 3.14.
    - Integer: Represents whole numbers, such as 42L.
    - Character: Represents text data, enclosed in quotes, like "Hello, World!"
    - Logical: Represents binary values, TRUE or FALSE.

2. Variables:
    - Use the assignment operator <- or = to assign values to variables. For example:

age <- 25
name <- "John"
is_student <- TRUE

3. Vectors:
    - Vectors are one-dimensional arrays that hold elements of the same data type.

heights <- c(175, 180, 165, 172)

4. Data Frames:
    - Data frames are two-dimensional data structures that resemble tables. They can store data of different data types.

data_frame <- data.frame(Name = c("Alice", "Bob", "Charlie"),
            Age = c(25, 30, 22),
            Score = c(90, 85, 88))

5. Basic Operations:
    - You can perform arithmetic operations on numeric data.

x <- 5
y <- 3
sum <- x + y
*product <- x * y*

6. Functions:
   - R has numerous built-in functions for various tasks, such as mean(), sum(), length(), and str() (for data frame structure).

7. Data Import:
   - To import data into R, you can use functions like read.csv(), read.table(), or specialized packages like readr or data.table.

my_data <- read.csv("data.csv")

8. Data Exploration:
   - Use functions like head(), tail(), and summary() to explore your data.

head(my_data)
summary(my_data)

9. Data Selection and Filtering:
   - You can select specific columns or filter rows based on conditions using square brackets [ ].

selected_column <- my_data$Column_Name
filtered_data <- my_data[my_data$Age > 25,]

10. Data Visualization:
    - Use libraries like ggplot2 for creating visualizations like scatterplots, histograms, and bar charts.

library(ggplot2)
ggplot(data=my_data, aes(x=Age, y=Score)) +
  geom_point()

11. Conditional Statements:
    - Use if, else if, and else for conditional logic.

if (condition) {
  # Code to execute if condition is true
} else if (another_condition) {
  # Code to execute if another_condition is true
} else {
  # Code to execute if no conditions are true
}

12. Loops:
    - R supports for and while loops for repetitive tasks.

```
for (i in 1:10) {
  print(i)
}
```

These are some of the basic concepts and commands you'll need to start with R programming for data analysis. As you become more comfortable with these fundamentals, you can explore more advanced topics and libraries to perform more complex data analyses.

R packages and libraries for sports analytics

There are several R packages and libraries that are specifically designed for sports analytics or can be useful in sports-related data analysis. Here are some of the key R packages and libraries for sports analytics:

1. SportsAnalytics:
   - The "SportsAnalytics" package provides access to datasets and functions for sports analytics. It covers various sports, including basketball, baseball, and football, and includes data related to player performance, team statistics, and game results.

install.packages("SportsAnalytics")
library(SportsAnalytics)

2. NBA_api:
   - The "NBA_api" package allows you to retrieve NBA data using the NBA's official API. You can access player statistics, team information, and game results.

devtools::install_github("jasoncaruso/NBA_api")
library(NBA_api)

3. nflscrapR:
   - "nflscrapR" is an R package for analyzing NFL data. It provides access to detailed play-by-play data, player statistics, and game results.

devtools::install_github("nflverse/nflscrapR")
library(nflscrapR)

4. SportsDataIO:
   - The "SportsDataIO" package allows you to access sports data from various sources, including sports betting odds, team statistics, and player information.

install.packages("SportsDataIO")
library(SportsDataIO)

5. GGally:
- "GGally" is an extension of the "ggplot2" package and provides additional functions for creating interactive and exploratory visualizations. It can be useful for creating advanced sports-related visualizations.

install.packages("GGally")
library(GGally)

6. tidyr and dplyr:
- These are part of the "tidyverse" collection of packages and are commonly used for data wrangling and manipulation. They are valuable for cleaning and preparing sports data for analysis.

install.packages("tidyverse")
library(tidyr)
library(dplyr)

7. plotly:
- "plotly" is an interactive graphing library that allows you to create interactive and dynamic sports visualizations. It's particularly useful for creating interactive dashboards.

install.packages("plotly")
library(plotly)

8. xgboost and caret:
- "xgboost" and "caret" are machine learning packages often used for predictive modeling in sports analytics. They can help build models for game outcome prediction, player performance, and injury risk assessment.

install.packages("xgboost")
library(xgboost)
install.packages("caret")
library(caret)

9. sportsdatadownloader:
- The "sportsdatadownloader" package provides tools to scrape sports data from various websites and sources. It's useful for collecting data for custom sports analytics projects.

devtools::install_github("0xdata/sportsdatadownloader")
library(sportsdatadownloader)

These R packages and libraries cater to a wide range of sports analytics needs, from data collection and preparation to visualization and modeling. Depending on your specific sports analytics project or area of interest, you can choose the appropriate packages to enhance your analysis and gain valuable insights.

Chapter 14: Advanced Topics in R for Sports Analytics

Advanced data manipulation in R for Sports Analytics

Advanced data manipulation in R is essential for sports analytics because sports datasets often require complex transformations and aggregations to extract meaningful insights. Here are some advanced data manipulation techniques in R that are particularly useful for sports analytics:

1. Reshaping Data with tidyr:
    - The tidyr package in the tidyverse is essential for reshaping data. You can use functions like gather() and spread() to convert data between wide and long formats, which is often necessary for various analyses.

library(tidyr)

 # Convert data from wide to long format
long_data <- gather(wide_data, key = "variable", value = "value", -id)

 # Convert data from long to wide format
wide_data <- spread(long_data, key = "variable", value = "value")

2. Data Aggregation with dplyr:
    - The dplyr package is incredibly powerful for data aggregation and manipulation. You can use functions like group_by(), summarize(), and mutate() to create custom variables, calculate aggregates, and group data by specific criteria.

library(dplyr)

 # Group data by team and calculate average points scored
team_stats <- sports_data %>%
  group_by(Team) %>%
  summarise(Avg_Points = mean(Points))

 # Create a new variable based on existing data
sports_data <- sports_data %>%

    mutate(Points_Per_Game = Points / Games_Played)

3. Data Joining with dplyr:
   - Use the join functions in dplyr (e.g., left_join(), inner_join()) to combine multiple datasets based on common columns. This is useful when working with data from multiple sources.

# Join player data with team data using common Team_ID
player_team_data <- left_join(player_data, team_data, by = "Team_ID")

4. Lag and Lead Functions:
   - In sports analytics, it's often necessary to calculate metrics that depend on previous or future game results or player performance. The lag() and lead() functions in dplyr can help with this.

library(dplyr)

# Calculate the difference in scores between the current and previous game
sports_data <- sports_data %>%
  arrange(Team, Game_Date) %>%
  group_by(Team) %>%
  mutate(Score_Diff_Prev_Game = Points - lag(Points))

5. Window Functions:
   - Window functions like cumsum(), cummean(), and cummax() allow you to calculate cumulative statistics over a specific window of data points. These can be valuable for tracking trends and streaks in sports analytics.

library(dplyr)

# Calculate cumulative points scored by a player over the season
player_data <- player_data %>%
  arrange(Player_ID, Game_Date) %>%
  group_by(Player_ID) %>%
  mutate(Cumulative_Points = cumsum(Points))

6. Conditional Data Manipulation:
   - Use conditional statements and functions like ifelse() and case_when() to apply different transformations to your data based on specific conditions. This can be helpful when dealing with outlier detection or data cleaning.

library(dplyr)

Replace negative values with NA in a variable

```
sports_data <- sports_data %>%
  mutate(Score = ifelse(Score < 0, NA, Score))
```

7. Looping and Iteration:
   - While vectorized operations are preferred in R, there may be cases where looping is necessary, such as iterating through multiple datasets or performing custom calculations.

```
for (i in 1:length(data_list)) {
  data_list[[i]] <- data_list[[i]] %>%
    mutate(New_Variable = some_custom_function(Old_Variable))
}
```

These advanced data manipulation techniques in R, combined with the power of the tidyverse packages, will allow you to handle complex sports datasets and perform in-depth analyses for sports analytics. Always keep in mind the specific requirements of your analysis and adapt these techniques accordingly.

Building custom functions and packages for Sports Analytics in R

Building custom functions and packages in R for sports analytics can help streamline your workflow, ensure code reusability, and contribute to the R community's knowledge. Here's a step-by-step guide on how to create custom functions and packages for sports analytics in R:

Creating Custom Functions:

    Identify the Need:
   - Determine the specific tasks or operations in your sports analytics workflow that could benefit from custom functions. These could include data cleaning, feature engineering, or statistical analysis.

    Write the Function:
   - Use R's function creation syntax to write your custom function. Here's a basic template:

```
custom_function <- function(arg1, arg2, ...) {
  # Function code
  result <- ...
  return(result)
```

}

Test the Function:
- Before packaging it, thoroughly test your custom function on sample data to ensure it works as intended. Address any bugs or issues that arise during testing.

Document the Function:
- Add comments and documentation to your function to describe its purpose, arguments, and expected outputs. Use the roxygen2 package for formal documentation.

```r
#' Custom Function for Calculating Player Performance
#'
#' This function calculates player performance based on certain criteria.
#'
#' @param data A data frame containing player data.
#' @param criteria A vector of criteria to be considered.
#' @return A data frame with player performance scores.
#'
#' @examples
#' custom_function(data = player_data, criteria = c("points", "assists"))
#'
custom_function <- function(data, criteria) {
  # Function code
}
```

Package Your Functions:
- While you can use custom functions directly in your R scripts, it's often beneficial to package them for better organization and sharing.

Creating an R Package:

Install and Load devtools:
- You'll need the devtools package to create and manage your R package. Install it if you haven't already:

```r
install.packages("devtools")
library(devtools)
```

Create Package Directory:
- Use create_package() to set up the directory structure for your package:

```r
create_package("my_sports_package")
```

Add Functions:
- Place your custom functions in the R/ directory within your package directory.

Document Functions:
- Add documentation to your functions using roxygen2-style comments as shown earlier.

Build and Load the Package:
- Use load_all() from devtools to load your package for development purposes:

load_all("path_to_your_package_directory")

Check the Package:
- Ensure your package adheres to best practices and has no errors using devtools::check():

devtools::check()

Install the Package Locally:
- You can install your package locally using devtools::install():

devtools::install("path_to_your_package_directory")

Test the Package:
- Create test scripts in the tests/ directory of your package to thoroughly test the functions.

Document the Package:
- Write a package documentation file (DESCRIPTION) that includes metadata about your package.

Share and Publish:
- Share your package with the R community by publishing it on CRAN or another repository if desired.

Once your custom functions and package are created and tested, you can use them in your sports analytics projects by loading the package using library(my_sports_package) and calling your custom functions. Creating custom functions and packages in R promotes code reusability and maintainability, making it easier to tackle complex sports analytics tasks efficiently.

Integrating R with other tools and languages for Sports Analytics

Integrating R with other tools and languages can enhance sports analytics capabilities, especially when different tools offer specialized features. Here are ways to integrate R with other tools and languages for sports analytics:

1. Python:
   - Python and R are two of the most popular languages for data analysis. You can integrate them using packages like reticulate, which allows you to run Python code within R scripts and vice versa. This is particularly useful when you want to leverage Python's machine learning libraries like scikit-learn for predictive modeling.

 # Load the reticulate package
library(reticulate)

 # Use Python code within R
py_code <- "print('Hello from Python')"
py_run_string(py_code)

2. SQL Databases:
   - To work with large sports datasets, integrate R with SQL databases using packages like DBI and RSQLite. This allows you to query and analyze data stored in relational databases directly from R.

library(DBI)
library(RSQLite)

 # Connect to a SQLite database
con <- dbConnect(RSQLite::SQLite(), "my_database.db")

 # Execute SQL query
*query <- "SELECT * FROM player_stats WHERE Points > 20"*
results <- dbGetQuery(con, query)

 # Close the database connection
dbDisconnect(con)

3. Excel:
   - You can integrate R with Microsoft Excel using packages like readxl to import data from Excel files and writexl to export R data frames to Excel format. This

is useful for sharing results and collaborating with stakeholders who use Excel.

library(readxl)
library(writexl)

 # Import data from Excel
my_data <- read_excel("my_data.xlsx")

 # Export data frame to Excel
write_xlsx(my_data, "output_data.xlsx")

4. Tableau:
   - Tableau is a popular data visualization tool. You can integrate R scripts into Tableau using calculated fields and R integration options. This enables you to create advanced visualizations and analytics dashboards combining R's analytical power with Tableau's visualization capabilities.

5. Power BI:
   - Similar to Tableau, Microsoft Power BI allows integration with R. You can use R scripts to perform custom data transformations and analyses within Power BI reports and dashboards.

6. APIs and Web Services:
   - Access sports data from APIs and web services using R's packages like httr or curl. You can retrieve real-time game scores, player statistics, and other sports-related information.

7. GIS Tools:
   - For sports analytics involving location data, integrate R with Geographic Information System (GIS) tools like QGIS or ArcGIS using the sf package. This allows you to analyze and visualize spatial data.

8. Sports-Specific Software:
   - Some sports analytics software like Sportscode, OptaPro, and SportVU provide APIs or data export options that can be used to import data into R for further analysis.

9. Jupyter Notebooks:
   - Combine the strengths of R and Python by using Jupyter notebooks that support both languages. Jupyter notebooks are particularly popular for data exploration and visualization in sports analytics.

10. Machine Learning Frameworks:
- Integrate R with machine learning frameworks like TensorFlow or PyTorch using packages like `reticulate` or `tensorflow` to build and deploy machine learning models for sports analytics tasks.

Integrating R with other tools and languages provides flexibility and allows you to leverage the strengths of each tool for different aspects of sports analytics, from data manipulation and analysis to visualization and modeling. Choose the integration methods that best suit your specific sports analytics project requirements and tools you're comfortable with.

Tips for efficient and scalable R coding for Sports Analytics

Efficiency and scalability are crucial when working on sports analytics projects in R, especially when dealing with large datasets and complex analyses. Here are some tips for writing efficient and scalable R code for sports analytics:

1. Use Data Frames and Tidy Data Principles:
   - Store your data in data frames and follow tidy data principles (each variable is a column, each observation is a row) to make data manipulation and analysis more efficient.

2. Limit Data Loading:
   - Load only the necessary data into memory to reduce overhead. Use functions like readr::read_csv() with the col_types argument to specify column types and avoid unnecessary conversions.

3. Optimize Data Cleaning:
   - Use vectorized operations and functions like dplyr and tidyr for data cleaning. Minimize the use of loops, which can be slow in R.

4. Parallelize Computations:
   - Utilize parallel computing to distribute computations across multiple cores or nodes. Packages like parallel and foreach can help speed up tasks like bootstrapping or cross-validation.

5. Profiling and Benchmarking:
   - Profile your code using tools like profvis and profiler to identify bottlenecks. Use the microbenchmark package to benchmark different code implementations and choose the fastest one.

6. Efficient Data Aggregation:
   - When aggregating data, use data.table or dplyr for efficient grouping and summarization. Take advantage of database backends like dtplyr when working with large datasets.

7. Indexing and Subsetting:
   - Optimize data indexing and subsetting. Use the data.table package for fast and memory-efficient subsetting operations.

8. Avoid Unnecessary Copying:
   - Be mindful of data copying. Many R functions create copies of data frames by default. Use dplyr functions with the mutate() family to modify data in place whenever possible.

9. Memory Management:
   - Monitor and manage memory usage. Functions like gc() (garbage collection) can help free up memory, especially when working with large datasets.

10. R Packages for Large Datasets:
- Consider using packages like `ff`, `fst`, or `disk.frame` for working with datasets that are too large to fit into memory. These packages allow you to efficiently process and analyze data stored on disk.

11. Proficient Use of Apply Functions:
- Use `lapply()`, `sapply()`, and `vapply()` instead of `for` loops when applying functions to elements of a list or a vector. These functions are more efficient.

12. Regular Expressions:
- When working with text data (e.g., player names, team names), use regular expressions (regex) efficiently to extract or manipulate information. The `stringr` package is helpful for this.

13. Optimize Visualization:
- Use packages like `ggplot2` for visualization, but be mindful of rendering time for complex plots. Consider aggregating data for large datasets to speed up plotting.

14. Modular Code:
- Write modular and reusable code. Break your code into functions and scripts, and organize your project structure logically.

15. Documentation and Comments:
- Document your code thoroughly using comments, and provide clear explanations of your logic and assumptions. This helps not only you but also others who may collaborate on the project.

16. Version Control:
- Use version control systems like Git and platforms like GitHub to track changes in your code, collaborate with others, and manage different versions of your analysis.

17. Avoid Global Variables:
- Minimize the use of global variables, as they can lead to unintended side effects. Use functions with explicit input and output parameters instead.

18. Memory Profiling:
- Use memory profiling tools like `profmem` to identify memory-intensive parts of your code and optimize them.

19. Benchmarking:
- Regularly benchmark your code to track performance improvements or regressions as you make changes.

20. Keep Learning:
- Stay up-to-date with R and data science best practices. New packages and techniques are continually emerging that can improve your efficiency.

By following these tips and continuously optimizing your R code, you can work more efficiently and scale your sports analytics projects to handle larger datasets and more complex analyses.

Chapter 15: Conclusion and Practical Applications

Summarizing key takeaways from the book of Sports Analytics in R

While I don't have access to a specific book titled "Sports Analytics in R," I can provide you with some key takeaways that you might find in such a book. These takeaways encompass fundamental concepts and principles often covered in sports analytics using R:

- Data is the Foundation: Sports analytics heavily relies on data collection, management, and analysis. A robust understanding of data sources, data cleaning, and preprocessing is essential.
- R as a Powerful Tool: R is a versatile programming language for sports analytics due to its extensive data manipulation, statistical analysis, and visualization capabilities. It offers packages like dplyr, ggplot2, and tidyr that streamline analysis.
- Importance of Data Visualization: Data visualization in R, particularly using packages like ggplot2, helps communicate insights effectively. Visualizations like scatterplots, heatmaps, and boxplots can reveal patterns and trends in sports data.
- Statistical Analysis: R provides a wide range of statistical techniques to analyze sports data, including regression analysis, hypothesis testing, and time series analysis. These methods are crucial for understanding player performance and team dynamics.
- Machine Learning and Predictive Modeling: Machine learning algorithms in R, such as decision trees, random forests, and support vector machines, are valuable for predicting game outcomes, player performance, and injury risk.
- Custom Metrics: Building custom metrics and indices tailored to specific sports and analysis goals is common in sports analytics. R allows for the creation of these metrics.
- Team and Player Evaluation: Team-level and player-level performance metrics are essential for assessing and comparing performance across different sports. R facilitates the creation of these metrics.
- Game Simulation and Strategy Optimization: Sports analytics often involves simulating games or optimizing team strategies to maximize success. R can be used to build simulation models and perform optimization tasks.

- Injury Analysis and Prevention: Injury prediction models and strategies for injury prevention are vital in sports analytics. R can be used to develop and validate these models.
- Predictive Modeling for Drafts and Recruitment: R can help sports organizations make informed decisions during drafts and recruitments by building predictive models to assess player potential.
- Fan Engagement and Marketing: Analyzing fan behavior and preferences using R can inform marketing strategies and enhance the fan experience.
- Ethical Considerations: Ethical issues, data privacy, and fair play are crucial aspects of sports analytics that must be considered throughout the analysis process.
- Emerging Technologies: Staying updated on emerging technologies like AI and machine learning is essential, as they continue to impact sports analytics.
- Challenges and Failures: Sports analytics projects often face challenges and may not always succeed. Learning from failures and adjusting strategies is important for growth.
- Efficiency and Scalability: Efficient coding practices and scalability are vital when working with large sports datasets, and R offers tools and techniques to address these needs.

Encouraging further exploration in sports analytics

Encouraging further exploration in sports analytics can be both rewarding and intellectually stimulating. Whether you are a beginner or a seasoned analyst, here are some tips to foster your curiosity and drive for diving deeper into this exciting field:

Stay Informed:
- Keep up-to-date with the latest trends and developments in sports analytics. Follow sports analytics blogs, websites, and journals to stay informed about new research, techniques, and applications.

Participate in Online Communities:
- Join online sports analytics communities and forums, such as Reddit's r/sportsanalytics or the Sports Analytics World Series community, where you can ask questions, share insights, and learn from other enthusiasts.

Attend Conferences and Workshops:
- Consider attending sports analytics conferences and workshops, such as the MIT Sloan Sports Analytics Conference. These events provide

opportunities to network with experts, attend informative sessions, and learn about cutting-edge research.

Online Courses and MOOCs:
- Enroll in online courses and MOOCs (Massive Open Online Courses) related to sports analytics. Platforms like Coursera, edX, and Udemy offer courses on various sports analytics topics.

Learn Advanced Tools and Languages:
- Explore advanced tools and languages that complement R, such as Python, SQL, or machine learning frameworks like TensorFlow and PyTorch. These skills can broaden your analytical capabilities.

Collaborate on Projects:
- Collaborate with others on sports analytics projects. Join or create teams to work on data analysis, predictive modeling, or visualization projects. Collaboration fosters learning and creativity.

Read Books and Research Papers:
- Dive into books and research papers on sports analytics. Look for publications by prominent authors in the field. This deep dive can provide valuable insights and research methodologies.

Build a Portfolio:
- Create a portfolio of sports analytics projects to showcase your skills and knowledge. Share your analyses and visualizations on platforms like GitHub or personal blogs.

Leverage Open Data Sources:
- Explore open data sources related to sports, such as the NFL's Big Data Bowl or the NBA's open data initiative. These datasets can serve as a foundation for your analyses.

Experiment with Different Sports:
- Don't limit yourself to a single sport. Explore analytics in various sports, as each may present unique challenges and opportunities for analysis.

Challenge Yourself:
- Set challenging goals and projects. Tackle complex problems, such as injury prediction, draft strategy, or game simulations, to push the boundaries of your skills.

Stay Ethical and Responsible:

- Maintain ethical standards in your work. Respect data privacy, integrity, and fairness, especially when dealing with sensitive sports-related information.

Seek Mentorship:
- If possible, find a mentor in the field of sports analytics who can provide guidance, share insights, and offer career advice.

Teach Others:
- Teaching is an excellent way to solidify your own understanding. Consider mentoring or teaching sports analytics to others, whether through formal classes, workshops, or online tutorials.

Stay Curious and Persistent:
- Sports analytics is a continuously evolving field. Stay curious, be persistent, and don't be discouraged by challenges or setbacks. Each obstacle is an opportunity to learn and grow.

By embracing these tips and maintaining a passion for sports analytics, you can explore new horizons, make valuable contributions to the field, and enjoy the exciting journey of uncovering hidden insights in the world of sports.

Real-world applications of sports analytics using R

Sports analytics using R has found numerous real-world applications across various sports. Here are some notable examples:

Player Performance Analysis:
- Teams use R to analyze player performance statistics, such as scoring efficiency, field goal percentages, and player tracking data (e.g., distance covered, speed) to make informed decisions about player rotations and strategies.

Injury Prediction and Prevention:
- Sports organizations employ R to build injury prediction models. By analyzing player health data and injury history, they can identify risk factors and develop strategies to prevent injuries and optimize player health.

Draft and Recruitment Strategies:
- R is used to assess the potential of draft prospects and free agents. Analytics teams analyze player statistics, physical attributes, and

combine performance data to make informed draft selections and recruitment decisions.

Game Strategy Optimization:
- Coaches and analysts use R to simulate and optimize game strategies. They can simulate various scenarios, such as different lineup combinations or offensive and defensive strategies, to find the most effective approach.

Performance Metrics for Teams and Players:
- R is used to create custom performance metrics that go beyond traditional statistics. Teams and analysts develop advanced metrics to evaluate player and team performance accurately.

Scouting and Talent Identification:
- Scouting departments use R to process and analyze vast amounts of scouting reports, video footage, and player data to identify emerging talent and potential recruits.

Fan Engagement and Marketing:
- Sports teams and leagues analyze fan behavior using R. They segment fan data, track engagement metrics, and optimize marketing strategies to enhance fan experiences and increase fan engagement.

Game Outcome Prediction:
- Analysts use R to build predictive models for game outcomes. By considering team strengths, player statistics, and historical performance, these models help sportsbooks and fans make predictions.

Social Media Sentiment Analysis:
- Sports organizations leverage R to perform sentiment analysis on social media data. They track fan sentiment and engagement, helping them gauge public opinion and make data-driven decisions.

Performance Improvement:
- R is used in player development programs. Coaches and trainers analyze player data to identify areas for improvement, tailor training regimens, and track progress over time.

Referee and Umpire Analysis:
- Sports leagues use R to analyze referee or umpire performance. By reviewing calls and game events, they assess the accuracy and consistency of officiating.

Salary Cap Management:
- R is employed for salary cap management in leagues like the NBA and NFL. Teams use analytics to optimize roster construction while staying within budget constraints.

Ticket Pricing and Sales Optimization:
- Sports teams use R to determine ticket pricing strategies. They analyze historical sales data, fan demographics, and game factors to set ticket prices that maximize revenue.

Health and Fitness Monitoring:
- Athletes' health and fitness data are tracked using wearables and sensors. R is used to analyze this data to monitor athletes' performance, recovery, and overall well-being.

Performance Analysis in Esports:
- Esports teams use R to analyze in-game data, such as player movements and actions, to gain insights into team strategies and individual player performance.

These real-world applications of sports analytics using R demonstrate the wide-ranging impact of data-driven decision-making in the world of sports. R's flexibility and analytical capabilities make it a valuable tool for sports organizations, coaches, analysts, and fans looking to gain a competitive edge and a deeper understanding of the games they love.

Final thoughts and future directions

In the world of sports analytics, the journey is ongoing, and the future holds exciting possibilities. Here are some final thoughts and future directions for sports analytics:

Integration of Advanced Technologies: The sports industry will continue to integrate advanced technologies such as computer vision, sensors, and wearable devices to collect high-frequency data for more precise player and game analysis.

Machine Learning and AI: Machine learning and artificial intelligence will play an increasingly significant role in sports analytics, with more sophisticated predictive models, game simulations, and player performance analysis.

Player Welfare and Health Analytics: The emphasis on player welfare will grow, with teams investing heavily in injury prevention and player health analytics. This will include the use of biometrics, genomics, and personalized training programs.

Fan Engagement and Immersive Experiences: Teams and leagues will leverage analytics to enhance fan engagement further. Virtual reality, augmented reality, and personalized content will provide fans with immersive experiences.

Global Expansion of Analytics: Sports analytics will continue to expand globally, not just in major professional sports but also in collegiate, amateur, and emerging sports markets.

Data Privacy and Ethics: As data collection intensifies, ethical considerations and data privacy will be critical. Organizations will need to establish guidelines and protocols to protect player and fan data.

Interdisciplinary Collaboration: Sports analytics will benefit from increased collaboration between data scientists, sports scientists, medical professionals, and coaches. Interdisciplinary teams will drive innovation in player performance and injury prevention.

Esports Analytics: Esports analytics will continue to evolve rapidly as competitive video gaming gains mainstream attention. The analysis of esports data will become more sophisticated, providing valuable insights for teams and players.

Real-time Analytics: Real-time analytics will become the norm, allowing coaches to make instant decisions based on live data feeds during games.

Custom Metrics and Model Interpretability: There will be a focus on developing custom metrics and enhancing the interpretability of complex machine learning models to ensure that insights are actionable and easily understood by coaches and players.

Education and Skill Development: Education programs and courses in sports analytics will become more prevalent, nurturing a new generation of analysts and professionals in the field.

Quantified Self for Athletes: Athletes at all levels will embrace the "quantified self" concept, using wearable devices and analytics to optimize their personal performance and health.

Global Sports Analytics Community: A global community of sports analysts and enthusiasts will continue to grow, sharing knowledge, insights, and best practices.

Ethical Considerations: The ethical implications of sports analytics will be a central theme, with ongoing discussions about fairness, bias, and responsible use of data.

In conclusion, the world of sports analytics is a dynamic and rapidly evolving field with a promising future. It will continue to transform how sports are played, coached, and experienced by athletes and fans alike. As technology and data analytics capabilities advance, the opportunities for innovation and discovery in sports analytics are limitless, making it an exciting space to watch and be a part of.

Example of Sports Analytics using R Code

Here's a large example of sports analytics using R code. In this example, we'll focus on analyzing NBA player performance data, including data collection, data preprocessing, exploratory data analysis (EDA), and the creation of custom metrics. We'll use the tidyverse package for data manipulation and visualization.

```r
# Load required libraries
library(tidyverse)

# Load NBA player performance data (example dataset)
nba_data <- read.csv("nba_player_data.csv")

# Data Preprocessing
# Filter out rows with missing values
nba_data <- na.omit(nba_data)

# Create a new variable for player efficiency rating (PER)
nba_data <- nba_data %>%
   mutate(PER = (points + rebounds + assists + steals + blocks - turnovers) / games_played)

# Explore Data
# Summary statistics
summary(nba_data)

# Visualize player age distribution
ggplot(nba_data, aes(x = age)) +
  geom_histogram(binwidth = 1, fill = "skyblue", color = "black") +
  labs(title = "NBA Player Age Distribution", x = "Age", y = "Frequency")

# Visualize the relationship between points and assists
ggplot(nba_data, aes(x = points, y = assists)) +
  geom_point(color = "blue") +
  geom_smooth(method = "lm", se = FALSE, color = "red") +
  labs(title = "Scatter Plot of Points vs. Assists",
       x = "Points",
       y = "Assists")

# Create a subset of top-performing players (PER > 20)
top_players <- nba_data %>%
  filter(PER > 20)
```

```r
# Visualize the top players' points distribution
ggplot(top_players, aes(x = points, fill = position)) +
  geom_density(alpha = 0.5) +
  labs(title = "Distribution of Points for Top Players (PER > 20)",
    x = "Points",
    y = "Density") +
  facet_wrap(~position)

# Export the top players' data to a new CSV file
write.csv(top_players, "top_nba_players.csv")

# Perform linear regression to predict points based on assists
lm_model <- lm(points ~ assists, data = nba_data)

# Summarize the linear regression model
summary(lm_model)

# Make predictions
new_data <- data.frame(assists = c(5, 10, 15))
predictions <- predict(lm_model, newdata = new_data)

# Display the predictions
print(predictions)
```

In this example, we start by loading NBA player performance data from a CSV file.

We preprocess the data, create a custom metric (Player Efficiency Rating or PER), and perform exploratory data analysis (EDA) by visualizing player age distributions, points vs. assists relationships, and more.

We then filter the dataset to identify top-performing players (PER > 20) and visualize the distribution of points for these players, categorized by their positions.

Finally, we export the data of top players to a new CSV file, fit a linear regression model to predict points based on assists, and make predictions for new data points.

Advanced example of Soccer Analytics with R code

Here's a more complex and advanced example of sports analytics using R code. In this example, we will analyze soccer (football) match data, including data collection, data preprocessing, advanced statistical modeling, and visualization. We'll use the tidyverse, lubridate, and ggplot2 packages, among others.

```r
# Load required libraries
library(tidyverse)
library(lubridate)

# Load soccer match data (example dataset)
soccer_data <- read.csv("soccer_match_data.csv")

# Data Preprocessing
# Convert date column to a date object
soccer_data$date <- as.Date(soccer_data$date)

# Calculate match duration in minutes
soccer_data <- soccer_data %>%
   mutate(match_duration = as.numeric(difftime(end_time, start_time, units = "mins")))

# Calculate goals per minute for each team
soccer_data <- soccer_data %>%
  group_by(home_team, away_team) %>%
  summarise(goals_per_minute = sum(goals_scored) / sum(match_duration))

# Advanced Statistical Analysis
# Fit a Poisson regression model to predict goals scored
poisson_model <- glm(goals_scored ~ home_team + away_team, data = soccer_data, family = "poisson")

# Summarize the Poisson regression model
summary(poisson_model)

# Visualize the distribution of goals per minute
ggplot(soccer_data, aes(x = goals_per_minute, fill = home_team)) +
  geom_density(alpha = 0.5) +
   labs(title = "Distribution of Goals Per Minute",
      x = "Goals Per Minute",
```

```
        y = "Density") +
  facet_wrap(~home_team)

# Predict goals for an upcoming match
new_match <- data.frame(home_team = "TeamA", away_team = "TeamB")
predicted_goals <- predict(poisson_model, newdata = new_match, type = "response")

# Display the predicted goals
print(predicted_goals)
```

In this advanced example:

> We load soccer match data and preprocess it, converting date columns and calculating match duration in minutes.

> We calculate the goals scored per minute for each team in each match, providing a more nuanced measure of team performance.

> We fit a Poisson regression model to predict the number of goals scored by each team in a match, considering the home and away teams as predictors.

> We visualize the distribution of goals scored per minute, comparing home teams and away teams.

> Finally, we use the Poisson regression model to predict the number of goals for an upcoming match between "TeamA" and "TeamB."

This example demonstrates more advanced techniques, including statistical modeling and visualization, commonly used in sports analytics to gain deeper insights into team performance and predict match outcomes.

Advanced example of NBA analytics with R code

Here's a complex and advanced example of NBA (National Basketball Association) analytics using R code. In this example, we will analyze player performance data, create custom metrics, conduct advanced statistical modeling, and visualize the results. We'll use the tidyverse, lubridate, and other packages.

```r
# Load required libraries
library(tidyverse)
library(lubridate)

# Load NBA player performance data (example dataset)
nba_data <- read.csv("nba_player_data.csv")

# Data Preprocessing
# Convert date column to a date object
nba_data$date <- as.Date(nba_data$date)

# Calculate player age
nba_data <- nba_data %>%
  mutate(age = floor(as.numeric(difftime(Sys.Date(), birthdate, units = "days")) / 365.25))

# Create a new variable for player efficiency rating (PER)
nba_data <- nba_data %>%
  mutate(PER = (points + rebounds + assists + steals + blocks - turnovers) / games_played)

# Advanced Statistical Analysis
# Fit a linear regression model to predict player PER based on various stats
lm_model <- lm(PER ~ points + rebounds + assists + steals + blocks + turnovers, data = nba_data)

# Summarize the linear regression model
summary(lm_model)

# Create a subset of top-performing players (PER > 20)
top_players <- nba_data %>%
  filter(PER > 20)

# Visualize the relationship between points and PER for top players
ggplot(top_players, aes(x = points, y = PER)) +
  geom_point(color = "blue") +
```

```
geom_smooth(method = "lm", se = FALSE, color = "red") +
labs(title = "Scatter Plot of Points vs. PER for Top Players",
    x = "Points",
    y = "PER")

# Export the top players' data to a new CSV file
write.csv(top_players, "top_nba_players.csv")

# Predict player PER for a new player based on their statistics
new_player_stats <- data.frame(points = 20, rebounds = 5, assists = 7, steals = 2, blocks = 1, turnovers = 3)
predicted_PER <- predict(lm_model, newdata = new_player_stats)

# Display the predicted PER
print(predicted_PER)
```

In this advanced NBA analytics example:

> We load NBA player performance data and preprocess it, including converting date columns, calculating player age, and creating a custom metric (Player Efficiency Rating or PER).
>
> We fit a linear regression model to predict player PER based on various statistics, including points, rebounds, assists, steals, blocks, and turnovers.
>
> We summarize the linear regression model to understand the relationships between performance metrics and PER.
>
> We create a subset of top-performing players (PER > 20) and visualize the relationship between points and PER for these players.
>
> We export the data of top-performing players to a new CSV file.
> Finally, we use the linear regression model to predict the PER of a new player with specific performance statistics.

This example showcases advanced statistical modeling and visualization techniques commonly used in NBA analytics to evaluate player performance and predict player outcomes.

Advanced example of NFL analytics with R code

Here's an advanced example of NFL (National Football League) analytics using R code. In this example, we'll analyze NFL player performance data, conduct advanced statistical modeling, and visualize the results. We'll use the tidyverse, lubridate, and other packages.

```r
# Load required libraries
library(tidyverse)
library(lubridate)

# Load NFL player performance data (example dataset)
nfl_data <- read.csv("nfl_player_data.csv")

# Data Preprocessing
# Convert date column to a date object
nfl_data$date <- as.Date(nfl_data$date)

# Calculate player age
nfl_data <- nfl_data %>%
  mutate(age = floor(as.numeric(difftime(Sys.Date(), birthdate, units = "days")) / 365.25))

# Advanced Statistical Analysis
# Fit a logistic regression model to predict player Pro Bowl selection
logistic_model <- glm(Pro_Bowl ~ age + passing_yards + rushing_yards + receptions + tackles + sacks,
            data = nfl_data, family = "binomial")

# Summarize the logistic regression model
summary(logistic_model)

# Create a subset of Pro Bowl selected players
pro_bowl_players <- nfl_data %>%
  filter(predict(logistic_model, type = "response") > 0.5)

# Visualize the distribution of passing yards for Pro Bowl players
ggplot(pro_bowl_players, aes(x = passing_yards)) +
  geom_density(fill = "blue", alpha = 0.5) +
  labs(title = "Distribution of Passing Yards for Pro Bowl Players",
     x = "Passing Yards",
     y = "Density")
```

```
# Export the Pro Bowl players' data to a new CSV file
write.csv(pro_bowl_players, "pro_bowl_players.csv")

# Predict Pro Bowl selection for a new player based on their stats
new_player_stats <- data.frame(age = 25, passing_yards = 4000, rushing_yards
= 800, receptions = 60, tackles = 70, sacks = 10)
predicted_pro_bowl <- predict(logistic_model, newdata = new_player_stats,
type = "response")

# Display the predicted Pro Bowl selection probability
print(predicted_pro_bowl)
```

In this advanced NFL analytics example:

> We load NFL player performance data and preprocess it, including converting date columns, calculating player age, and cleaning the dataset.

> We fit a logistic regression model to predict player Pro Bowl selection based on player age, passing yards, rushing yards, receptions, tackles, and sacks. Pro Bowl selection is treated as a binary outcome (1 for selected, 0 for not selected).

> We summarize the logistic regression model to understand the relationships between player statistics and Pro Bowl selection.

> We create a subset of Pro Bowl selected players based on the model's predictions.

> We visualize the distribution of passing yards for Pro Bowl selected players.

> We export the data of Pro Bowl selected players to a new CSV file.

> Finally, we use the logistic regression model to predict the probability of Pro Bowl selection for a new player with specific performance statistics.

This example demonstrates advanced statistical modeling and visualization techniques commonly used in NFL analytics to evaluate player performance and predict player outcomes, such as Pro Bowl selection.

Advanced example of Baseball analytics with R code

Here's an advanced example of baseball analytics using R code. In this example, we'll analyze baseball player performance data, conduct advanced statistical modeling, and visualize the results. We'll use the tidyverse, lubridate, and other packages.

```r
# Load required libraries
library(tidyverse)
library(lubridate)

# Load baseball player performance data (example dataset)
baseball_data <- read.csv("baseball_player_data.csv")

# Data Preprocessing
# Convert date column to a date object
baseball_data$date <- as.Date(baseball_data$date)

# Calculate player age
baseball_data <- baseball_data %>%
  mutate(age = floor(as.numeric(difftime(Sys.Date(), birthdate, units = "days")) / 365.25))

# Advanced Statistical Analysis
# Fit a linear regression model to predict player batting average (BA) based on various stats
lm_model <- lm(BA ~ hits + walks + strikeouts + home_runs + stolen_bases, data = baseball_data)

# Summarize the linear regression model
summary(lm_model)

# Create a subset of high-performing players (BA > 0.300)
high_ba_players <- baseball_data %>%
  filter(BA > 0.300)

# Visualize the relationship between hits and batting average (BA) for high-performing players
ggplot(high_ba_players, aes(x = hits, y = BA)) +
  geom_point(color = "blue") +
  geom_smooth(method = "lm", se = FALSE, color = "red") +
```

```r
  labs(title = "Scatter Plot of Hits vs. BA for High-Performing Players",
       x = "Hits",
       y = "BA")

# Export the high-performing players' data to a new CSV file
write.csv(high_ba_players, "high_ba_players.csv")

# Predict player BA for a new player based on their stats
new_player_stats <- data.frame(hits = 150, walks = 60, strikeouts = 80, home_runs = 20, stolen_bases = 10)
predicted_BA <- predict(lm_model, newdata = new_player_stats)

# Display the predicted batting average (BA)
print(predicted_BA)
```

In this advanced baseball analytics example:

> We load baseball player performance data and preprocess it, including converting date columns, calculating player age, and cleaning the dataset.
>
> We fit a linear regression model to predict player batting average (BA) based on various statistics, including hits, walks, strikeouts, home runs, and stolen bases.
>
> We summarize the linear regression model to understand the relationships between performance metrics and batting average.
>
> We create a subset of high-performing players based on a threshold batting average of 0.300.
>
> We visualize the relationship between hits and batting average (BA) for high-performing players.
>
> We export the data of high-performing players to a new CSV file.
> Finally, we use the linear regression model to predict the batting average (BA) for a new player with specific performance statistics.

This example showcases advanced statistical modeling and visualization techniques commonly used in baseball analytics to evaluate player performance and predict batting average.

Advanced example of Hockey analytics with R code

Here's an advanced example of hockey analytics using R code. In this example, we'll analyze hockey player performance data, conduct advanced statistical modeling, and visualize the results. We'll use the tidyverse, lubridate, and other packages.

```
# Load required libraries
library(tidyverse)
library(lubridate)

# Load hockey player performance data (example dataset)
hockey_data <- read.csv("hockey_player_data.csv")

# Data Preprocessing
# Convert date column to a date object
hockey_data$date <- as.Date(hockey_data$date)

# Calculate player age
hockey_data <- hockey_data %>%
  mutate(age = floor(as.numeric(difftime(Sys.Date(), birthdate, units = "days")) / 365.25))

# Advanced Statistical Analysis
# Fit a mixed-effects model to predict player goals scored based on various stats
library(lme4)
mixed_model <- lmer(goals_scored ~ assists + penalty_minutes + (1 | team), data = hockey_data)

# Summarize the mixed-effects model
summary(mixed_model)

# Create a subset of high-scoring players (goals scored > 30)
high_scoring_players <- hockey_data %>%
  filter(goals_scored > 30)

# Visualize the relationship between assists and goals scored for high-scoring players
ggplot(high_scoring_players, aes(x = assists, y = goals_scored)) +
  geom_point(color = "blue") +
  geom_smooth(method = "lm", se = FALSE, color = "red") +
```

```
    labs(title = "Scatter Plot of Assists vs. Goals Scored for High-Scoring Players",
         x = "Assists",
         y = "Goals Scored")

# Export the high-scoring players' data to a new CSV file
write.csv(high_scoring_players, "high_scoring_players.csv")

# Predict player goals scored for a new player based on their stats
new_player_stats <- data.frame(assists = 40, penalty_minutes = 20)
predicted_goals <- predict(mixed_model, newdata = new_player_stats, re.form = NA)

# Display the predicted goals scored
print(predicted_goals)
```

In this advanced hockey analytics example:

> We load hockey player performance data and preprocess it, including converting date columns, calculating player age, and cleaning the dataset.

> We fit a mixed-effects model to predict player goals scored based on various statistics, including assists, penalty minutes, and account for team-specific effects using random intercepts.

> We summarize the mixed-effects model to understand the relationships between performance metrics and goals scored.

> We create a subset of high-scoring players based on a threshold of 30 goals scored.

> We visualize the relationship between assists and goals scored for high-scoring players.

> We export the data of high-scoring players to a new CSV file.
> Finally, we use the mixed-effects model to predict the goals scored for a new player with specific performance statistics.

This example showcases advanced statistical modeling and visualization techniques commonly used in hockey analytics to evaluate player performance and predict goal-scoring outcomes.

Advanced example of Tennis analytics with R code

Here's an advanced example of tennis analytics using R code. In this example, we'll analyze tennis player performance data, conduct advanced statistical modeling, and visualize the results. We'll use the tidyverse, lubridate, and other packages.

```
# Load required libraries
library(tidyverse)
library(lubridate)

# Load tennis player performance data (example dataset)
tennis_data <- read.csv("tennis_player_data.csv")

# Data Preprocessing
# Convert date column to a date object
tennis_data$date <- as.Date(tennis_data$date)

# Calculate player age
tennis_data <- tennis_data %>%
  mutate(age = floor(as.numeric(difftime(Sys.Date(), birthdate, units = "days")) / 365.25))

# Advanced Statistical Analysis
# Fit a logistic regression model to predict player match wins
logistic_model <- glm(match_wins ~ age + double_faults + aces + break_points_won,
            data = tennis_data, family = "binomial")

# Summarize the logistic regression model
summary(logistic_model)

# Create a subset of high-performing players (match wins > 20)
high_winning_players <- tennis_data %>%
  filter(match_wins > 20)

# Visualize the relationship between age and match wins for high-performing players
ggplot(high_winning_players, aes(x = age, y = match_wins)) +
  geom_point(color = "blue") +
  geom_smooth(method = "lm", se = FALSE, color = "red") +
  labs(title = "Scatter Plot of Age vs. Match Wins for High-Performing Players",
       x = "Age",
       y = "Match Wins")
```

```r
# Export the high-performing players' data to a new CSV file
write.csv(high_winning_players, "high_winning_players.csv")

# Predict player match wins for a new player based on their stats
new_player_stats <- data.frame(age = 28, double_faults = 10, aces = 50, break_points_won = 30)
predicted_wins <- predict(logistic_model, newdata = new_player_stats, type = "response")

# Display the predicted match wins probability
print(predicted_wins)
```

In this advanced tennis analytics example:

> We load tennis player performance data and preprocess it, including converting date columns, calculating player age, and cleaning the dataset.

> We fit a logistic regression model to predict player match wins based on various statistics, including age, double faults, aces, and break points won. Match wins are treated as a binary outcome (1 for wins, 0 for losses).

> We summarize the logistic regression model to understand the relationships between player statistics and match wins.

> We create a subset of high-performing players based on a threshold of 20 match wins.

> We visualize the relationship between age and match wins for high-performing players.

> We export the data of high-performing players to a new CSV file.

> Finally, we use the logistic regression model to predict the probability of match wins for a new player with specific performance statistics.

This example showcases advanced statistical modeling and visualization techniques commonly used in tennis analytics to evaluate player performance and predict match outcomes.

Advanced example of Golf analytics with R code

Here's an advanced example of golf analytics using R code. In this example, we'll analyze golf player performance data, conduct advanced statistical modeling, and visualize the results. We'll use the tidyverse and other packages.

```r
# Load required libraries
library(tidyverse)

# Load golf player performance data (example dataset)
golf_data <- read.csv("golf_player_data.csv")

# Data Preprocessing
# Calculate the average score for each player
golf_data <- golf_data %>%
   mutate(average_score = (score_round1 + score_round2 + score_round3 + score_round4) / 4)

# Advanced Statistical Analysis
# Fit a linear regression model to predict player earnings based on their average score
lm_model <- lm(earnings ~ average_score, data = golf_data)

# Summarize the linear regression model
summary(lm_model)

# Create a subset of top-earning players (earnings > 1000000)
top_earning_players <- golf_data %>%
  filter(earnings > 1000000)

# Visualize the relationship between average score and earnings for top-earning players
ggplot(top_earning_players, aes(x = average_score, y = earnings)) +
  geom_point(color = "blue") +
  geom_smooth(method = "lm", se = FALSE, color = "red") +
   labs(title = "Scatter Plot of Average Score vs. Earnings for Top-Earning Golfers",
     x = "Average Score",
     y = "Earnings")

# Export the top-earning players' data to a new CSV file
write.csv(top_earning_players, "top_earning_golfers.csv")
```

```r
# Predict player earnings for a new golfer based on their average score
new_player_stats <- data.frame(average_score = 70)
predicted_earnings <- predict(lm_model, newdata = new_player_stats)

# Display the predicted earnings
print(predicted_earnings)
```

In this advanced golf analytics example:

> We load golf player performance data and preprocess it by calculating the average score for each player based on their scores in multiple rounds.

> We fit a linear regression model to predict player earnings based on their average score. Earnings are treated as the dependent variable, and average score is the independent variable.

> We summarize the linear regression model to understand the relationship between average score and earnings.

> We create a subset of top-earning players based on a threshold of $1,000,000 in earnings.

> We visualize the relationship between average score and earnings for top-earning players.

> We export the data of top-earning players to a new CSV file.
> Finally, we use the linear regression model to predict the earnings for a new golfer with a specific average score.

This example showcases advanced statistical modeling and visualization techniques commonly used in golf analytics to evaluate player performance and predict earnings based on performance metrics like average score.

Advanced example of Boxing analytics with R code

In this example, we'll analyze boxing match outcomes and create visualizations to gain insights. We'll use the tidyverse package for data manipulation and visualization.

Please note that real-world boxing analytics often require more detailed data, including individual fighter statistics, and involve complex modeling techniques, which may not be covered comprehensively in this simplified example.

```r
# Load required libraries
library(tidyverse)

# Load hypothetical boxing match data (example dataset)
boxing_data <- read.csv("boxing_match_data.csv")

# Data Preprocessing
# Calculate the total number of matches for each boxer
boxing_data <- boxing_data %>%
  mutate(total_matches = wins + losses + draws)

# Calculate the winning percentage for each boxer
boxing_data <- boxing_data %>%
  mutate(win_percentage = wins / total_matches)

# Advanced Statistical Analysis
# Identify boxers with a winning percentage above 0.8
top_boxers <- boxing_data %>%
  filter(win_percentage > 0.8)

# Visualize the distribution of winning percentages for top boxers
ggplot(top_boxers, aes(x = win_percentage)) +
  geom_histogram(binwidth = 0.1, fill = "blue", color = "black") +
  labs(title = "Distribution of Winning Percentages for Top Boxers",
    x = "Winning Percentage",
    y = "Frequency")

# Export the data of top boxers to a new CSV file
write.csv(top_boxers, "top_boxers.csv")

# Predict the outcome of a hypothetical match between Boxer A and Boxer B
```

```r
# Here, we assume that Boxer A has a winning percentage of 0.85, and Boxer B has 0.75
predicted_winner <- ifelse(0.85 > 0.75, "Boxer A", "Boxer B")

# Display the predicted winner
print(paste("Predicted Winner:", predicted_winner))
```

In this advanced boxing analytics example:

> We load hypothetical boxing match data and preprocess it by calculating the total number of matches and the winning percentage for each boxer.
>
> We identify top boxers with a winning percentage above 0.8.
>
> We visualize the distribution of winning percentages for top boxers using a histogram.
>
> We export the data of top boxers to a new CSV file.
>
> Finally, we predict the outcome of a hypothetical match between two boxers based on their winning percentages.

Please keep in mind that this example is simplified and hypothetical. In real-world boxing analytics, you would typically work with more detailed fighter statistics and apply more advanced modeling techniques to make predictions and gain insights into the sport.

Printed in Dunstable, United Kingdom